MECHANISMS OF LYMPHOCYTE ACTIVATION AND IMMUNE REGULATION IV

Cellular Communications

ADVANCES IN EXPERIMENTAL MEDICINE AND BIOLOGY

Recent Volumes in this Series

MECHANISMS OF LYMPHOCYTE ACTIVATION AND IMMUNE REGULATION IV

Cellular Communications

Edited by

Sudhir Gupta

University of California
Irvine, California

and

Thomas A. Waldmann

National Cancer Institute
National Institutes of Health
Bethesda, Maryland

SPRINGER SCIENCE+BUSINESS MEDIA, LLC

Library of Congress Cataloging-in-Publication Data

Mechanisms of lymphocyte activation and immune regulation IV :
  cellular communications / edited by Sudhir Gupta and Thomas A.
  Waldmann.
       p.   cm. -- (Advances in experimental medicine and biology ;
  323)
       "Proceedings of the Fourth International Conference on Lymphocyte
  Activation and Immune Regulation, held February 14-16, 1992, in
  Newport Beach, California"--T.p. verso.
     Includes bibliographical references and index.
     ISBN 978-0-306-44312-1   ISBN 978-1-4615-3396-2 (eBook)
     DOI 10.1007/978-1-4615-3396-2
     1. Lymphocyte transformation--Congresses. 2. Immune response-
  -Regulation--Congresses. 3. Cellular signal transduction-
  -Congresses.   I. Gupta, Sudhir. II. Waldmann, Thomas, 1930-   .
  III. International Conference on Lymphocyte Activation and Immune
  Regulation (4th : 1992 : Newport Beach, Calif.) IV. Series.
     [DNLM: 1. Lymphocyte Transformation--congresses. 2. Lymphocytes-
  -immunology--congresses. 3. Lymphocytes--physiology--congresses.]
  QR185.8.L9M43   1992
  616.07'9--dc20   92-26881
                                                                  CIP

Proceedings of the Fourth International Conference on Lymphocyte Activation
and Immune Regulation, held February 14-16, 1992, in Newport Beach, California

ISBN 978-0-306-44312-1

© 1992 Springer Science+Business Media New York
Originally published by Plenum Press, New York in 1992

PREFACE

In recent years rapid progress has been made in the areas of T cell and B cell biology, cell-cell and cell-matrix/stroma interactions. The use of isolated subunits of the T cell receptor invariant chains has been instrumental in defining their role in signal transduction and tyrosine phosphorylation. A role of *src* family phosphotyrosine kinases in T cell activation has been demonstrated and several phosphotyrosine kinase substrates have been identified and their functions characterized. Homologous recombinant techniques have led to the development of murine strains that lack CD4 or CD8 expression. These models are likely to be instrumental in studying the role of T cell subsets in autoimmune disorders, tissue transplant rejection and tumor rejection. A role of major histocompatibility complex I in the development of T cell subsets and NK cells has been defined. Recent data suggest a role of interaction between plasma membrane molecules of activated T helper cells and B cells, B cells primed with plasma membrane of activated T helper cells and cytokines, and interaction between bone marrow stromal cells and B cell progenitors and precursors, in the B cell development, proliferation, and differentiation. The structure and functions of adhesion molecules, especially with regard to signal transduction and homing events, are better defined. A role of adhesion molecules in the pathogenesis of human diseases is realized and therapeutic modalities directed against some of the adhesion molecules in the treatment of experimental models of certain human diseases have been initiated. The progress in the above areas was the main theme of the 4th International Conference on Lymphocyte Activation and Immune Regulation held in Newport Beach, California on February 14-16, 1992.

This book is divided into four sections. **Section I** deals with the structure of the T cell receptors and their role in signal transduction and tyrosine phosphorylation, role of *src* family of tyrosine kinases in T cell activation, identification and characterization of phosphotyrosine kinase substrates, and a role of CD45 and CD28 in signal transduction. This section also includes discussion regarding involvement of protein kinase C isoforms in the expression of P-glycoprotein, role of $\gamma\delta$ T cells in T cell development and in the pathogenesis of autoimmune diseases, and IL-2 receptors and IL-2 receptor directed therapy for autoimmune and malignant disorders. **Section II** includes discussions regarding a role of tolerance and major histocompatibility complex I in T cell development. Also included is the use of homologous recombinant technique and differential involvement of phosphotyrosine kinase $p56^{lck}$ and $p59^{fyn}$ in T cell development. The instructional and stochastic models of T cell differentiation are discussed. The development, activation, proliferation, and differentiation of B cells are the main focus of **Section III**. Included in this section are: properties and activities of T helper cell plasma membranes in effector phase of cognate help for B cell proliferation and differentiation; role of stromal cell interaction and IL-7 with B cell progenitors and precursor; and B cells to induce proliferation and differentiation. In addition, a role of CD23 in B cell activation is discussed. **Section IV** deals with various aspects of adhesion molecules, including a role in signal transduction, a functional role of the integrin α cytoplasmic tail, regulation of β_2 integrins, and the role of adhesion receptors in multiple leukocyte homing events.

This book should be of interest to basic immunologists and molecular biologists.

The Editors wish to thank Miss Nancy Doman for her excellent secretarial assistance.

Sudhir Gupta
Thomas A. Waldmann

CONTENTS

T CELL RECEPTOR STRUCTURE AND FUNCTION: ANALYSIS BY EXPRESSION OF PORTIONS OF ISOLATED SUBUNITS

Isaac Engel, Francois Letourneur, John T. B. Houston, Tom H. M. Ottenhoff, and Richard D. Klausner

Cell Biology and Metabolism Branch
National Institute of Child Health and Human Development
National Institutes of Health
Bethesda, MD 20892

A. FUNCTION OF INDIVIDUAL CD3 CHAINS AND ZETA DIMERS

The T cell antigen receptor (TCR) is a multi-component cell surface complex composed of the products of at least six genes (1,2). Specific recognition of antigen/MHC is mediated by two chains of the TCR complex (generally α and β) that are expressed as disulfide linked heterodimers and display a high degree of clonotypic diversity. Associated with this specialized heterodimer are the invariant dimers CD3 ε-δ and CD3 ε-γ and a disulfide-linked dimer formed by members of the ζ gene family (typically a ζ-ζ homodimer). The invariant chains of the TCR have been shown to be necessary for efficient assembly and transport of the receptor to the cell surface (3), and are also thought to be essential for signal transduction.

Upon stimulation of the TCR either after antigen/MHC recognition or cross-linking with anti-receptor antibody, the most proximal and rapid event is the activation of one or more tyrosine kinases (4). These processes result in tyrosine phosphorylation of endogenous substrates including phospholipase Cγ1 (5,6) or the ζ chain (7), ensuing further biochemical activating events. However, direct assessment of the function of individual CD3 chains and ζ dimers in the initiation of these signal transduction events is limited by the complexity of the assembly of the TCR (3). Strict architectural editing mechanisms prevent cell surface expression of partial TCR/CD3 complexes or individual chains. To overcome these analytical problems several models have been developed.

Mechanisms of Lymphocyte Activation and Immune Regulation IV: Cellular Communications
Edited by S. Gupta and T.A. Waldmann, Plenum Press, New York, 1992

The identification of ζ deficient T cells allowed our laboratory to begin to examine the function of this polypeptide (8). In the absence of the ζ chain, the remaining subunits of the receptor assemble in and leave the endoplasmic reticulum but are primarily targeted for lysosomal degradation. Introduction of the cDNA encoding the ζ chain restored full surface expression and function. Surface expression can be reconstituted with altered cDNAs encoding truncated forms of the protein, however cells expressing such receptors are markedly deficient in their ability to respond to antigen (9).

Further evidence supporting a role for ζ in T cell activation has been obtained by the analysis of chimeric proteins made between the external domains of either CD8 or CD4 and the ζ cytoplasmic tail (10,11). Cross-linking of CD8-ζ chimeras expressed in Jurkat T cells was shown to activate pathways normally associated with TCR triggering. In addition, CD4 chimeras containing the cytoplasmic tail of ζ, η or FcεRIγ when expressed in cytotoxic T cells can direct the killing of cells that possess targets for the extracellular domain of CD4. These experiments demonstrated that the ζ chain can initiate cell activation independent of the other chains of the TCR complex.

Recently we have reported on our use of similar chimeras containing the extracellular domain of the α chain of the human interleukin 2 receptor (Tac) and the cytoplasmic tail of either ζ or FcεRIγ (12). Zeta and FcεRIγ chimeras, when externally cross-linked, were effective at eliciting IL2 production in T cells as well as serotonin release in rat basophilic leukemia cells. The signaling potencies of ζ tails truncated at position 108 or deleted from position 66 to 114 suggested the presence of several functional domains in ζ. These data indicated that ζ may be the essential operative protein for signal transduction through the TCR. However, two observations suggested that ζ alone did not account for all the functional capacities of the TCR. First in the absence of any ζ protein, the small number of incomplete TCR complexes that are transported to the cell surface can signal when cross-linked. Second, TCR/CD3 complexes devoid of a functional ζ subunit (due to an extensive cytoplasmic deletion of ζ) are still able to trigger the production of IL-2 in response to antigen or super-antigen when expressed in a murine thymoma (13). We therefore used the chimeric subunit approach to assess the signaling and activation capacity of the cytoplasmic tail of the CD3 ε chain (14).

BW5147 cells deficient for α, β, CD3 δ and ζ chains were stably transfected with Tac-ζ and Tac-ε chimeric constructs, and the resulting transfectants were tested for IL-2 production in response to cross-linking with a monoclonal antibody to Tac. Tac-ζ transfectants were most effective at IL-2 production, but at sufficient doses of activating antibody, Tac-ε transfectants produced as much IL-2 as cells expressing Tac-ζ. Despite the fact that both ε and ζ cytoplasmic tails contain tyrosine residues, only Tac-ζ was tyrosine phosphorylated after cross-linking, while Tac-ε was highly serine phosphorylated. However, for transfectants expressing Tac-ε or Tac-ζ chimeras, multiple tyrosine phosphorylated substrates were apparent. Several substrates were shared but others were differentially phosphorylated. These differences suggest that the biochemical consequences of signaling through the ζ and ε tails are distinct and may be correlated with the activation of different tyrosine kinases. The presence of two polypeptides in a single receptor complex that can each induce cell activation raises interesting issues about the function of these chains in the context of a complete receptor.

In order to characterize further the domain in the cytoplasmic tail of CD3 ε responsible for the activation of tyrosine kinases, transfectants were established that express Tac-ε chimeras containing truncations, deletions and point mutations in the CD3-ε domain. The COOH terminal 22 amino-acid residues of ε were shown to be critical for IL-2 production. Within this domain two regions found to be particularly important. The first region included Tyr170, Glu171, Pro172 and Ile173. The second region included Asp179, Leu180, and Tyr181. Reduction of the spacing between these two critical regions destroyed function.

The domain critical for activation through Tac-CD3ε lies within a consensus sequence present in the cytoplasmic domains of components of the TCR, the B cell receptor and some Fc receptors (15). Furthermore, a similar intron/exon organization of this motif is shared between these different proteins, suggesting a possible evolutionary relationship. The signaling specificity of these related motifs can now be addressed using the chimeric model we have established.

B. HIGH EFFICIENCY EXPRESSION AND SOLUBILIZATION OF TCR αβ HETERODIMERS

A determination of the three-dimensional structure of the clonotypic TCR αβ heterodimer has been a goal of many laboratories interested in understanding the physical basis for the recognition of antigen/MHC complexes by T cells. The production of large amounts of a soluble form of the αβ heterodimer is necessary for X-ray crystallographic analysis. Furthermore, soluble heterodimer could also be useful for other structural studies, such as measuring the affinity between the heterodimer and antigen/MHC. However, there are a number of problems that complicate the isolation of soluble αβ heterodimer. The α and β chains are both type 1 transmembrane proteins, and as such are not soluble in the absence of detergent. Furthermore, the αβ heterodimer is not efficiently expressed at the cell surface unless associated with the other chains of the TCR complex (3). In addition, it is important to verify that any alterations made in the structure of the α and β chains in order to circumvent the problems of solubility and expression efficiency do not affect the heterodimer combining site.

Our laboratory has identified several sequence determinants in the TCR chains that act to cause retention or degradation of partial complexes (2,3). In particular, it has been found that the positive charges in the transmembrane domains of the α and β chains act to target these proteins for retention and/or degradation in the ER in the absence of the CD3 chains, and that there are also retention and degradation determinants in the extracellular domains of the α and β chains that are hidden by heterodimer formation. Based on this knowledge, it was predicted that αβ heterodimers would be efficiently expressed at the cell surface if the transmembrane domains of each chain were removed, or for example, replaced with sequences that target the chains for glycophosphatidyl inositol (GPI) anchoring, as has been reported by two groups (16,17). However, analysis of cells that express such constructs revealed that the transport of GPI-anchored α and β to the cell surface was extremely inefficient, with most of the protein remaining in the ER in a non-disulfide-linked form (18).

The above data suggested that the α and β chains do not form disulfide-linked dimers with high efficiency in the absence of the TCR complex invariant chains, and that disulfide-linked heterodimer formation may be critical for transport of αβ to the cell surface. We tested this hypothesis by replacing the native transmembrane and cytoplasmic domains of α and β with those of the TCR ζ chain. The ζ chain transmembrane domain has been shown to induce the formation of disulfide-linked ζ-ζ homodimers through a cysteine contained within the domain (19), and it also induces disulfide-linked dimerization when attached to heterologous proteins (12). We thus expected that chimeric α-ζ and β-ζ constructs would form disulfide-linked dimers with high efficiency. Because the ζ cytoplasmic domain has been shown to induce cell activation when attached to an extracellular domain that is directly cross-linked by antibody (10,11,12), cells that express α-ζ/β-ζ heterodimers at the cell surface could conceivably be activated by exposure to a specific antigen/MHC complex, and the structural integrity of the heterodimer could hence be confirmed. One disadvantage of using this approach towards efficient expression of αβ heterodimers is that type 1 transmembrane spanning proteins would be produced, and as such would not be suitable for crystallography. In order to circumvent this problem linker sequences containing the recognition site of the protease thrombin were inserted on the amino terminal side of the ζ transmembrane domain, such that digestion of the construct with thrombin would release soluble αβ heterodimer.

Chimeric constructs were made between TCR α and β cDNAs isolated from the T cell hybridoma 2B4 (20) and the murine TCR ζ cDNA using PCR cloning techniques. These constructs were stably transfected into the rat basophilic leukemia line RBL-2H3, a line that had been found in transient transfection experiments to express the constructs at the cell surface (21). Flow cytofluorometric analysis of stable transfectants revealed that cells transfected with both an α-ζ and a β-ζ chimera exhibited high cell surface levels of these polypeptides, while cells transfected with only one of the chimeric constructs had much lower levels of cell surface expression of the chimera (22). Biochemical analysis of the cell surface and intracellular forms of these chimeras confirmed that α-ζ/β-ζ disulfide-linked heterodimers were formed and transported to the cell surface with high efficiency, while disulfide-linked homodimers were formed with lower efficiency and exited the ER very slowly. Subsequent analysis of RBL-2H3 transfectants expressing native α and β chains, or α and β chains in which charged residues in the transmembrane domains were replaced with leucines in order to remove ER retention and degradation signals, determined that these constructs were predominantly retained in the ER in a non-disulfide-linked form. This demonstrated that the ζ domain was required for efficient disulfide-linked dimerization of the α-ζ and β-ζ chimeric proteins, and that the induction of α-ζ/β-ζ heterodimer formation allowed for efficient transport to the cell surface.

To test whether the antigen/MHC combining site of the α-ζ/β-ζ heterodimers expressed at the surface of RBL-2H3 transfectants was functional, serotonin release assays were performed. Transfectants were labeled with ³H-serotonin, and then incubated with Ek-expressing cells that were pre-pulsed with a peptide analog of moth cytochrome c, thus exposing the RBL-2H3 transfectants to an antigen/MHC complex recognized by the 2B4 TCR. Serotonin was released specifically upon exposure of the RBL-2H3 transfectants to cells presenting the appropriate antigen/MHC complex, thus demonstrating that αβ heterodimers on the cell surface of these transfectants are capable of recognizing antigen/MHC (22). These data also show that the αβ heterodimer can recognize antigen/MHC in the absence of the CD3 complex.

It was anticipated that the presence of a thrombin site in the linker sequence between the α or β domain and the ζ domain in each construct would allow the release of soluble αβ heterodimer upon digestion of α-ζ/β-ζ complexes with thrombin. Indeed it was found that α-ζ/β-ζ dimers immunoprecipitated from surface iodinated cells could be efficiently cleaved by thrombin (22). The protein released by thrombin treatment was found to be predominantly disulfide-linked and migrated under reducing and non-reducing conditions at positions consistent with that expected for a disulfide-linked αβ heterodimer truncated after the membrane-proximal cysteine residue. In addtion to the demonstration of thrombin cleavage, these data show that the ζ domain induces interchain disulfide-linkage through cysteine residues in the α and β domains as well as through the ζ transmembrane. Thrombin digestion of α-ζ/β-ζ protein that lacks thrombin sites yielded only minor amounts of proteolytic products, indicating that thrombin treatment does not result in significant degradation of the αβ heterodimer. We are now in the process of scaling up the harvest and digestion of α-ζ/β-ζ heterodimers in order to realize the production of large amounts of soluble αβ heterodimer.

The attachment of the ζ chain transmembrane and cytoplasmic domains to the α and β chain extracellular domains thus provides both for the efficient production and cell surface transport of αβ heterodimers as well as a means by which to test the structural integrity of the expressed heterodimer. Furthermore, the inclusion of protease cleavage sites between the extracellular and transmembrane domains of each construct allows the production of soluble αβ heterodimer that is likely to be very similar to the native form, which in turn should allow for a number of structural studies of the heterodimer, including crystallographic analysis and measurements of the affinity between TCRαβ, MHC, and antigen. We also anticipate that these soluble heterodimers could be used for the generation of anti-TCR antibodies that recognize native epitopes, and it is possible that specific heterodimers could be used therapeutically as well. Furthermore, it will be interesting to see if this general approach could be incorporated into strategies for the efficient production of other dimeric proteins.

REFERENCES

1. H.C. Clevers, B. Alarcon, T.E. Wileman, and C. Terhorst, The T-cell receptor/CD3 complex: A dynamic protein ensemble, *Annu. Rev. Immunol.* 6:629 (1988).

2. R.D. Klausner, J. Lippincott-Schwartz,and J.S. Bonifacino, The T-cell antigen receptor: Insights into organelle biology, *Annu. Rev. Cell. Biol.* 6:403 (1990).

3. R.D. Klausner, J. Lippincott-Schwartz,and J.S. Bonifacino, Architiectural editing: Regulating the surface expression of the multicomponent T-cell antigen receptor, *Current Topics in Membranes and Transport* 36:31 (1990).

(4) R.D. Klausner, and L.E. Samelson, T cell antigen receptor activation pathways: The tyrosine kinase connection, *Cell* 64:875 (1991).

(5) D.J. Park, H.W. Rho, and S.G. Rhee, CD3 stimulation causes phosphorylation of phospholipase C-γ1 on serine and tyrosine residues in a human T-cell line, *Proc. Natl. Acad. Sci. USA* 88:5453 (1991).

(6) A.Weiss, G. Koretsky, R. Schatzman, and T. Kadlecek, Functional activation of the T-cell antigen receptor induces tyrosine phosphorylation of phospholipase C-γ1, *Proc. Natl. Acad. Sci. USA* 88:5484 (1991).

(7) L.E. Samelson, M.D. Patel, A.M. Weissman, J.B. Harford, and R.D. Klausner, Antigen activation of murine T cells induces tyrosine phosphorylation of a polypeptide associated with the T cell antigen receptor, *Cell* 46:1083 (1986).

(8) J.J. Sussman et al., Failure to synthesize the T cell CD3-ζ chain: structure and function of a partial T cell receptor complex, *Cell* 52: 85 (1988).

(9) S.J. Frank et al., Structural mutations of the T cell receptor ζ chain and its role in T cell activation, *Science* 249:174 (1990).

(10) B.A. Irving and A. Weiss, The cytoplasmic domain of the T cell receptor ζ chain is sufficient to couple to receptor-associated signal transduction pathways, *Cell* 64:891 (1991).

(11) C. Romeo and B. Seed, Cellular immunity to HIV activated by CD4 fused to T cell or Fc receptor polypeptides, *Cell* 64:1037 (1991).

(12) F. Letourneur, and R.D. Klausner, T-cell and basophil activation through the cytoplasmic tail of T-cell-receptor ε family proteins, *Proc. Natl. Acad. Sci. USA* 88:8905 (1991).

(13) A.-M.K. Wegener et al., The T cell receptor/CD3 complex is composed of at least two autonomous transduction modules, *Cell* 68:83 (1992).

(14) F. Letourneur, and R.D. Klausner, Activation of T cells by a tyrosine kinase activation domain in the cytoplasmic tail of CD3 ε, *Science* 255:79 (1992).

(15) M. Reth, Antigen receptor tail clue, *Nature* 338, 383 (1989).

(16) A.Y. Lin et al., Expression of T cell antigen receptor heterodimers in a lipid-linked form, *Science* 249:677 (1990).

(17) A.E. Slanetz and A.L.M. Bothwell, Heterodimeric, disulfide-linked α/β T cell receptors in solution, *Eur. J. Immunol.* 21:179 (1991).

(18) C.K. Suzuki, and R.D. Klausner, unpublished observations.

(19) T.M. Rutledge, P. Cosson, N. Manolios, J.S. Bonifacino, and R.D. Klausner, in preparation.

(20) S. M. Hedrick et al., The fine specificity of antigen and Ia determinant recognition of T cell hybridoma clones specific for pigeon cytochrome c, *Cell* 30:141 (1982).

(21) I. Engel and R.D. Klausner, unpublished observations.

(22) I. Engel, T.H.M. Ottenhoff, and R.D. Klausner, High-efficiency expression and solubilization of functional T cell antigen receptor heterodimers, *Science* 256:1318.

THE T CELL ANTIGEN RECEPTOR TYROSINE KINASE PATHWAY

Lawrence E. Samelson, Mark Egerton, Pamela M. Thomas, and Ronald L. Wange

Cell Biology and Metabolism Branch
National Institute of Child Health and Human Development
National Institutes of Health
Bethesda, MD 20892

INTRODUCTION

It is now well accepted that T cell antigen receptor (TCR) engagement induces a cascade of protein phosphorylation with activation of both tyrosine and serine/threonine kinases. The most proximal events in this cascade appear to be the tyrosine phosphorylation of multiple intracellular proteins. This paper summarizes new data from our laboratory and places it in the context of observations made over the past several years. The initial problems that we and others have focused on are the identification and characterization of the kinases, phosphatases, and substrates that make up the elements of this pathway. With this information in hand, one can begin to understand the mechanism of signal transduction at a molecular level.

The first indication that the TCR is coupled to a protein tyrosine kinase (PTK) pathway was the observation that a 21 kD TCR-associated protein becomes tyrosine phosphorylated on TCR engagement[1]. It was later shown that this protein is the TCR-ζ chain, which becomes phosphorylated on tyrosine residues[2]. This receptor subunit has been intensively studied, but the

Mechanisms of Lymphocyte Activation and Immune Regulation IV: Cellular Communications
Edited by S. Gupta and T.A. Waldmann, Plenum Press, New York, 1992

9

function of the phosphorylation is not understood[3]. More recently a second PTK substrate in T cells has been identified. Studies on the kinetics of tyrosine phosphorylation and the effects of PTK inhibitors on T cell signalling pathways suggested that polyphosphoinositide breakdown in T cell is regulated by a PTK[4-6]. We postulated, based on work in growth factor systems, that phospholipase Cγ–1(PLC) itself is tyrosine phosphorylated and thus activated. Phosphorylation of this enzyme has been demonstrated by four laboratories[7-10]. The sites of tyrosine phosphorylation of T cell PLC are identical to those seen in fibroblasts strongly suggesting that these phosphorylations regulate enzyme activity.

CHARACTERIZATION OF NEW T CELL TYROSINE KINASE SUBSTRATES

There are a large number of tyrosine phosphorylated proteins that remain unidentified. One of the goals of our laboratory is to characterize these substrates and define their function. We have begun by purifying tyrosine phosphorylated proteins from lymph node T cells of MRL/*lpr* mice. These animals suffer from a genetic lymphoproliferative disorder, and large numbers of T cells can be isolated. These cells also overexpress the PTK p60*fyn*, and, perhaps as a consequence, contain a large number of constitutively tyrosine phosphorylated proteins[11,12]. The level of cellular tyrosine phosphorylation was further enhanced by treatment of the cells with two tyrosine phosphatase inhibitors, orthovanadate and phenylarsine oxide[13]. Immunoaffinity chromatography was then performed using antiphosphotyrosine monoclonal antibodies. Multiple proteins were isolated and the two most prominent, p81 and p100, were subjected to partial proteolysis and amino acid sequence analysis.

Peptide sequence from the 81kD substrate revealed that this protein is ezrin. This protein is known to be a substrate for growth factor receptor tyrosine kinases in epithelial cells. Addition of epidermal growth factor to A431 cells results in the tyrosine phosphorylation of 5-10%. In T cells we have confirmed that the 81kD substrate can be immunoprecipitated and immunoblotted with antibodies specific for ezrin. The protein is poorly tyrosine phosphorylated upon TCR engagement, but addition of orthovanadate and phenylarsine oxide results in a marked increase in tyrosine phosphorylation. Fifteen percent of ezrin is tyrosine phosphorylated in MRL/*lpr* T cells after phosphatase inhibitor treatment. Unlike in A431 cells ezrin remains in the cytosolic fraction. The function of ezrin and phosphorylated ezrin remains unclear.

Our interest in a 100kD PTK substrate originates in observations from earlier studies. This substrate is very rapidly tyrosine phosphorylated upon TCR engagement and its phosphorylation correlates with T cell activation. Peptides were prepared from purified p100 and the amino acid sequence was used to prepare oligonucleotide probes for screening a cDNA library prepared from MRL/*lpr* lymph node cells. Clones were obtained and characterized. The deduced amino acid sequence indicated that p100 is not a known PTK substrate. The predicted protein has a duplicated nucleotide binding domain and in this region is homologous to sec18, a yeast protein involved in intracellular vesicular transport. Antibodies have been generated against peptides made from the predicted sequence. These reagents have been used to demonstrate that the protein is expressed in T cells, B cells, mast cells and fibroblasts. Tyrosine phosphorylated p100 has been immunoprecipitated with anti-phosphotyrosine antibodies and immunoblotted with the anti-peptide antibodies. These studies have shown that the protein is a substrate for PTKs associated with the antigen receptor in T cells and B cells and with the IgE Fc receptor in mast cells. The level of tyrosine phosphorylation in some of these cells increases with receptor engagement, and addition of tyrosine phosphatase inhibitors results in a marked increase in tyrosine phosphate. The protein has been expressed in rat basophilic leukemia cells with an epitopic tag. Immunofluorescence microscopy shows that the protein has a cytoplasmic distribution. Future efforts will focus on attempts at determing the function of the protein and the effect of its tyrosine phosphorylation.

PROTEIN TYROSINE KINASES IMPLICATED IN T CELL ACTIVATION

Identification of the PTK(s) responsible for the tyrosine phosphorylation that is observed following TCR engagement is a major goal. Since the TCR itself lacks kinase domains it is clear that an associated kinases must be sought. The T cell specific src family kinase p56[lck] has been viewed as a major candidate. This kinase is asssociated with CD4 or CD8 and is activated when either of these surface molecules is crosslinked[15-17]. It appears that this kinase can be brought into the TCR complex when the the antigen-MHC complex is engaged[18,19]. It is likely that this kinase phosphorylates a subset of the T cell substrates.

A second member of the src family, p60[fyn] is expressed in T cells in a form unique to hematopoetic cells[20]. This kinase has been shown to be associated with the TCR[21]. This observation was made by extracting T cells in digitonin and testing immunoprecipitated TCR to determine whether a PTK co-precipitated. The presence of autophosphorylated p60[fyn] was detected in these immune complex kinase assays. More recently we have confirmed this observation by introducing water soluble homobifunctional chemical crosslinkers into permeabilized cells. Using these two techniques, digitonin extraction and chemical crosslinking, we have shown that 20-25% of the kinase is associated with the TCR. Conversely, however, only 2-4% of TCR is associated with p60[fyn]. Using the crosslinking protocol we have shown that the kinase and several associated proteins can be crosslinked to the TCR ζ chain. We do not know, however, the nature of this interaction since we can not demonstrate a direct p60[fyn]- TCR ζ association. The role of p60[fyn] is also not fully understood. Unlike p56[lck], there is as yet no evidence that p60[fyn] is activated by receptor crosslinking. However overexpression of the kinase leads to a T cell that is more sensitive to TCR engagement than a normal cell, suggesting that the kinase has some function in T cell activation[22].

An additional role for p60[fyn] is suggested by our observation that it is associated with the Thy-1 molecule in murine T cells. When Thy-1 is crosslinked with monoclonal antibodies, full T cell activation is induced[23]. The mechanism for Thy-1 mediated signalling is not fully understood. The molecule is not an integral membrane protein. Instead it is linked to the outer leaflet of the plasma membrane by a glycophosphatidylinositol(GPI) moiety. Earlier studies had indicated that full activation via Thy-1 crosslinking required TCR expression, and it was assumed that Thy-1 interacts with the TCR[24,25]. However studies from our laboratory indicated that in cells lacking

TCR, Thy-1 crosslinking still leads to PTK activation[26]. Our current data demonstrates that Thy-1 is associated with p60^{fyn}. In contrast to the requirement for digitonin or chemical crosslinking to show an association of the TCR with this kinase, the interaction of Thy-1 and p60^{fyn} is stable in Triton X-100. The association is detected in both murine hybridoma cells and thymocytes from young mice. About 25% of cellular p60^{fyn} can be detected with Thy-1. In contrast we detect only a small fraction of p56^{lck} associated with Thy-1. A third PTK present in T cells, p60^{yes}, does not co-precipitate with Thy-1. The association of this molecule with p60^{fyn} requires an intact GPI tail and it can be disrupted with the detergent octylglucoside. The latter result allows us to model our data based on results of Anderson et. al. and Brown and Rose who observe that GPI-linked proteins cluster in domains within the plasma membrane known as caveolae[27,28]. These structures and GPI-linked proteins, in particular, are poorly solubilized in Triton X-100, but can be completely solubilized in octylglucoside. The Thy-1-p60^{fyn} interaction may depend on glycolipid and protein interactions that define these microdomains. From a functional point of view the association suggests a means by which Thy-1 can be coupled to a PTK.

A third candidate for the PTK activated by the TCR has been discussed by Weiss and colleagues[29]. They demonstrated that when Jurkat T cells are stimulated by TCR engagement, a 70kD protein is tyrosine phosphorylated and associates with the TCR ζ chain. An activated PTK also associates with this chain. The TCRζ chain becomes tyrosine phosphorylated when an *in vitro* kinase assay is performed on anti-TCRζ complexes isolated from activated cells. Our studies of this system indicate that activation of the TCR by antibodies that bind either the TCRβ chain or the CD3ε chain results in association of the 70kD protein with both TCR ζ and CD3, and tyrosine phosphorylation of TCR ζ and CD3ε. The identity of the PTK responsible for the tyrosine phosphorylation is not certain, but some data point to the 70kD protein itself. Our studies with a photoreactive ATP analogue, [α^{32}P]-azidoanilido-ATP show that the 70kD protein can bind this reagent. The fact that this protein binds ATP is one piece of evidence that it iself is a kinase. Further studies are needed to fully understand the role of this 70kD molecule in signal transduction mediated by the TCR.

REFERENCES

1. L. E. Samelson, M. D. Patel, A. M. Weissman, J. B. Harford, and R. D. Klausner, Antigen activation of murine T cells induces tyrosine phosphorylation of a polypeptide associated with the T cell antigen receptor, *Cell* 46:1083 (1986).

2. M. Baniyash, P. Garcia-Morales, E. Luong, L. E. Samelson, and R. D. Klausner, The T cell antigen receptor ζ chain is tyrosine phosphorylated upon activation, *J. Biol. Chem.* 26: 18225 (1988).

3. S. J. Frank, I. Engel, T. M. Rutledge, and F. Letourneur, Structure/function analysis of the invariant subunits of the T cell antigen receptor, *Sem. Immunol.* 3:299 (1991).

4. C. H. June, M. C. Fletcher, J. A. Ledbetter, and L. E. Samelson, Increases in tyrosine phosphorylation are detectable before phospholipase C activation after T cell receptor stimulation, *J. Immunol.* 144:1591 (1990).

5. T. Mustelin, K. M. Coggeshall, N. Isakov, and A. Altman, T cell antigen receptor-mediated activation of phospholipase C requires tyrosine phosphorylation, *Science* 247:1584 (1990).

6. C. H. June, M. C. Fletcher, J. A. Ledbetter, G. L. Schieven, J. N. Siegel, A. F. Phillips, and L. E. Samelson, Inhibition of tyrosine phosphorylation prevents T cell receptor-mediated signal transduction, *Proc. Natl. Acad. Sci. USA.* 87:7722 (1990).

7. D. J. Park, H. W. Rho, and S. G. Rhee, CD3 stimulation causes phosphorylation of phospholipase C-γ1 on serine and tyrosine residues in a human T-cell line, *Proc. Natl. Acad. Sci. USA* 88:5453 (1991).

8. A. Weiss, G. Koretzky, R. Schatzman, and T. Kadlecek, Functional activation of the T cell antigen receptor induces tyrosine phosphorylation of phospholipase Cγ1,*Proc. Natl. Acad. Sci. USA* 88:5484 (1991).

9. J. P. Secrist, L. Karnitz, and R. T. Abraham, T-cell antigen receptor ligation induces tyrsoine phosphorylation of phospholipase C-γ1, *J. Biol. Chem.* 266:12135 (1991).

10. C. Granja, L.-L. Lin, E. J. Yunis, V. Relias, and J. D. Dasgupta, PLCγ1, a possible mediator of T cell receptor function, *J. Biol. Chem..* 266:16277 (1991).

11. L. E. Samelson, W. F. Davidson, H. C. Morse III, and R. D. Klausner, Abnormal constitutive tyrosine phosphorylation of the T cell antigen receptor in murine lymphoproliferative disorders, *Nature* 324:674 (1986).

12. T. Katagiri, J. P.-Y. Ting, R. Dy, C. Prokop, P. Cohen, and H. S. Earp, Tyrosine phosphorylation of a c-src-like protein is increased in membranes CD4-CD8- T lymphocytes from *lpr/lpr* mice, *Mol. Cell. Biol.* 9:4914 (1989).

13. P. Garcia-Morales, E. T. Luong, Y. Minami, R. D. Klausner, and L. E. Samelson, Tyrosine phosphorylation in T cells is regulated by phosphatase activity: studies with phenylarsine oxide, *Proc. Natl. Sci. USA.* 87:9255 (1990).

14. K. L. Gould, J. A. Cooper, A. Bretscher, and T. Hunter, The protein-tyrosine kinase substrate, p81, is homologous to a chicken microvillar core protein, *J. Cell Biol.* 102:660 (1986).

15. C. E. Rudd, J. M. Trevillyan, J. D. Dasgupta, L. L. Wong, and S. F. Schlossman, The CD4 receptor is complexed in detergent lysates to a protein-tyrosine kinase (pp58) from human T lymphocytes, *Proc. Natl. Acad. Sci. USA* 85:5190 (1988).

16. A. Veillette, M. A. Bookman, E. M. Horak, and J. B. Bolen, The CD4 and CD8 T cell surface antigens are associated with the internal membrane tyrosine kinase p56[lck], *Cell* 55:301 (1988).

17. A. Veillette, M. A. Bookman, E. M. Horak, L. E. Samelson, and J. B. Bolen, Signal transduction through the T-lymphocyte CD4 receptor involves the activation of the internal membrane tyrosine-protein tyrosine kinase p56[lck], *Nature* 338:257 (1989).

18. K. Saizawa, J. Rojo, and C. A. Janeway, Evidence for a physical association of CD4 and the CD3 α:β T cell receptor, *Nature* 328:260 (1987).

19. A. Kupfer, S. J. Singer, C. A. Janeway, and S. L. Swain, Coclustering of CD4 (L3T4) molecule with the T-cell receptor is induced by specific direct interaction of helper T cells and antigen-presenting cells, *Proc. Natl. Acad. Sci. USA* 84:5888 (1987).

20. M. P. Cooke, and R. M. Perlmutter, Expression of a novel form of the *fyn* proto-oncogene in hematopoietic cells, *The New Biologist* 1:66 (1990).

21. L. E. Samelson, A. F. Phillips, E. T. Luong, and R. D. Klausner, Association of the fyn protein tyrosine kinase with the T cell antigen receptor, *Proc. Natl. Acad. Sci. USA* 87:4358 (1990).

22. M. P. Cooke, K. M. Abraham, K. A. Forbush, and R. M. Perlmutter, Regulation of T cell receptor signaling by a src family protein-tyrosine kinase, *Cell* 65:281 (1991).

23. K. C. Gunter, T. R. Malek, and E. M. Shevach, T cell activating properties of an anti-Thy-1 monoclonal antibody: possible analogy to OKT3/LEU-4, *J. Exp. Med.* 159:716 (1984).

24. A. M. Schmitt-Verhulst, A. Guimezanes, C. Boyer, M. Poenie, R. Tsien, M. Buferne, C. Hoa, and L. Leserman, Pleotropic loss of activation pathways in a T-cell receptor α-chain deletion variant of cytolytic T-cell clone, *Nature* 325:628 (1987).

25. J. J. Sussman, T. Saito, E. M. Shevach, R. N. Germain, and J. D. Ashwell, Thy-1- and Ly-6-mediated lymphokine production and growth inhibition of a T cell hybridoma require co-expression of the T cell antigen receptor complex, *J Immunol.* 140:2520 (1988).

26. E. D. Hsi, J. N. Siegel, Y. Minami, E. T. Luong, R. D. Klausner, and L. E. Samelson, T cell activation induces rapid tyrosine phosphorylation of a limited number of cellular substrates, *J. Biol. Chem.* 264:10836 (1989).

27. R. G. W. Anderson, B. A. Kamen, K. G. Rothberg, and S. W. Lacey, Potocytosis: sequestration and transport of small molecules by caveolae, *Science* 255:410 (1992).

28. D. Brown, and J. K. Rose, Sorting of GPI-anchored proteins to glycolipid-enriched membrane subdomains during transport to the apical cell surface, *Cell* 68:533 (1992).

29. A. C. Chan, B. A. Irving, J. D. Fraser, and A. Weiss, The ζ chain is associated with a tyrosine kinase and upon T-cell antigen receptor stimulation associates with ZAP-70, a 70kDa tyrosine phosphoprotein, *Proc. Natl. Acad. Sci. USA* 88:9166 (1991).

DISSECTION OF THE Hb(64-76) DETERMINANT REVEALS THAT THE T CELL RECEPTOR MAY HAVE THE CAPACITY TO DIFFERENTIALLY SIGNAL

Brian D. Evavold and Paul M. Allen

Department of Pathology
Washington University School of Medicine
660 S. Euclid
St. Louis, MO 63110

Introduction

The T cell inducing determinant Hb(64-76) was initially identified in studies examining the processing and presentation of the self antigen hemoglobin[1]. Murine hemoglobin exists in two allelic forms, Hbb^s and Hbb^d. Immunization of H-2^k mice expressing the Hbb^s allele, e.g. CE/J, with purified hemoglobin from Hbb^d mice, e.g. CBA/J, resulted in a strong T cell response. This allelic response is entirely directed against a single determinant composed of residues 64-76 of the Hbb^{d}minor chain. Three of the 13 amino acid differences between the s and d allelic forms of hemoglobin are found in this region.

T cell hybridomas and clones were generated which recognized the Hb(64-76) determinant to be used as functional probes to detect the presence of self Hb/Ia complexes in vivo[1-3]. These studies have been previously reviewed in detail and will not be discussed here. In addition to being used as functional probes, these T cells have also been used to characterize in detail the Hb(64-76) determinant as a foreign antigen. Subsequent to the initial studies we have now generated a panel of both Th1 and Th2 clones which recognize the same determinant, Hb(64-76)/I-E^k[4]. In this report we will outline the characterization of the Hb(64-76) determinant using this panel of T cell clones and hybridomas, and how how this fine structure analysis revealed that the T cell receptor may have the capacity to differentially signal.

Dissection of the Hb(64-76) Determinant

The Hb(64-76) determinant was completely dissected by testing peptides containing single amino acid substitutions. The overall goal was to identify the

Mechanisms of Lymphocyte Activation and Immune Regulation IV: Cellular Communications
Edited by S. Gupta and T.A. Waldmann, Plenum Press, New York, 1992

17

amino acid residues of the determinant which were critical for contacting the T cell receptor and those which were important in binding to the I-E[k] molecule. This strategy has been used previously by our laboratory and others to analyze many different T cell inducing determinants[5-7]. Single conservative amino acid substitutions were made at each position of the Hb(64-76) determinant, and each peptide was tested for its ability to stimulate the panel of T cell hybridomas and Th1 and Th2 T cell clones. The results of these studies are summarized in Figure 1.

SUMMARY OF Hb(64–76) CRITICAL RESIDUES

		64	65	66	67	68	69	70	71	72	73	74	75	76
		G	K	K	V	I	T	A	F	N	E	G	L	K
	AK10.5	-	-	-	-	-	-	-	★	★	★	★	-	★
	BO1.10	-	-	-	-	-	★	-	★	★	-	-	★	-
T Cell Hybridomas	LK10.8	-	-	-	-	-	★	-	★	★	-	-	-	-
	OB12.6	-	-	-	-	-	-	-	-	★	-	-	-	-
	YO1.6	-	-	-	-	-	-	-	★	★	★	-	-	★
	PL.17	-	-	-	-	-	-	★	-	★	★	-	-	★
Th1 Clones	#1	-	-	-	-	★	-	★	★	★	★	-	★	-
	#2	-	-	-	-	-	★	★	★	★	★	★	-	★
	3.L2	-	-	-	-	-	★	-	★	★	★	-	-	★
	2.102	-	-	-	-	-	★	-	-	★	★	★	-	-
Th2 clones	EW5X	-	-	-	-	-	★	★	-	★	★	-	-	-

Figure 1. The critical residues of the Hb(64-76) determinant are shown. A ★ indicates that the residue is a critical one for the corresponding T cell. A "-" indicates that the position is not a critical residue. The following one letter amino acid codes are used: G=Gly, glycine, K=Lys, lysine, V=Val, valine, I=Ile, isoleucine, T=Thr, threonine, A=Ala, alanine, F=Phe, phenylalanine, N=Asn, asparagine, E=Glu, glutamic acid, L=Leu, leucine.

From these studies several conclusions were drawn[4]. The response to Hb(64-76) was oligoclonal in nature with no two T cells having an identical pattern of critical residues; however, several common features of the determinant were observed. The asparagine at position 72 (Asn-72) was the critical T cell contact residue for this determinant. For every T cell tested, Asn-72 was a critical residue. T cell contact with the Asn-72 residue was shown by two different methods. In direct binding studies to purified I-E[k] molecules, it was shown that the Gln-72 substituted peptide bound identically to the wild type sequence; despite its lack of ability to stimulate any of the T cells. Also, when bulk populations of T cells were primed with the Gln-72 peptide, a unique population of T cells was stimulated which did not cross-react with the wild type response. Taken together, these results indicate that Asn-72 is the major T cell contact residue of the Hb(64-76). In addition to the Asn-72 there are other secondary T cell contact residues, the main one being Glu-73. This residue was also shown by similar methods to be a T cell contact residue for 8 out of the 11 T cells. Interestingly, the amino acids in the allelic forms of Hb differed at both positions 72 and 73. Thus, the two residues play an important role in T cell recognition of Hb peptide. The third amino acid residue which was critical for most of the different T cells was Phe-71. By direct binding studies, this residue was found to be a critical residue for binding to the I-E^k molecule, in that the Phe-71 peptide bound 100-fold weaker than the wild type peptide. A graphical summary of the critical residues of Hb(64-76) is shown in Figure 2.

TCR AND MHC CONTACT RESIDUES OF Hb(64-76)

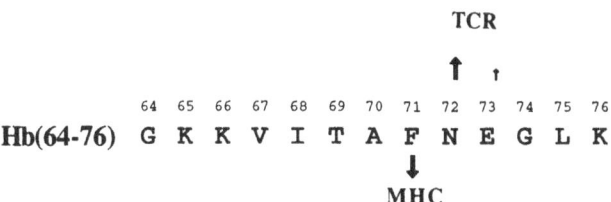

Figure 2. The T cell receptor(TCR) and I-E^k(MHC) contact residues of Hb(64-76) are shown. The size of the arrow indicates that relative importance of the residue in binding. Thus, the N-72 is the critical TCR contact residue, whereas the E-73 is a secondary T cell contact residue. There are other residues involved in contacting the TCR and I-E^k, but only the major contact residues are shown.

It is interesting to note that despite the common requirement for these 3 key residues, the number of critical residues varies dramatically between different T cells. For example, the OB12.6 T cell hybridoma only identified the Asn-72 as a critical residue, where in comparison, the Th1 clone, #2, has 7 critical residues. The molecular basis for these differences is not known, but it may be a reflection of the relative affinity of the T cell receptors for their ligand. From this dissection, it is now clear that T cells recognize the Hb(64-76)/I-E^k complex by mainly interacting with the Asn-72 sidechain. In addition, depending upon which T cell is being examined, other amino acid residues are also being recognized by the T cells as secondary contact residues, the main one being Glu-73.

Separation of lymphokine production from proliferation

In the above mentioned studies, the critical amino acids were determined for the T cell clones, by testing their ability to stimulate the T cell clones to proliferate. As proliferation of T cells is a distal event in T cell activation, we examined the response of T cells to the substituted peptides using more proximal assays for T cell activation. We chose to examine the production of IL-4 by the Th2 clone, 2.102, and from these studies we were able to separate IL-4 production from proliferation[8]. When each of the peptides were tested for their ability to induce IL-4 production and proliferation, three different patterns resulted. For some substituted peptides such as Gln-72, no IL-4 or proliferation occurred. Conversely, for other peptides such as Arg-76, both IL-4 and proliferation was observed. For one peptide, Asp-73, a very interesting observation was made. The Asp-73 peptide did not cause any proliferation of the Th2 T cell clone, 2.102 at concentrations even as high as 3mM; however, IL-4 was produced in amounts similar to that induced by the wild type peptide(Figure 3).

	64	65	66	67	68	69	70	71	72	73	74	75	76	T Cell Proliferation	T cell Lymphokine (IL-4)	B Cell Help
Hb(64-76)	G	K	K	V	I	T	A	F	N	E	G	L	K	+	+	+
Gln72	-	-	-	-	-	-	-	-	Q	-	-	-	-	-	-	-
Asp73	-	-	-	-	-	-	-	-	-	D	-	-	-	-	+	+
Arg76	-	-	-	-	-	-	-	-	-	-	-	-	R	+	+	+

Figure 3. The summary of the responses of the Th2 clone, 2.102, to substituted Hb(64-76) peptides. The peptide designation is shown on the left, and the sequence of the peptide is given using the one letter codes. A "-" indicates identity between the substituted peptide and the wild type Hb(64-76) sequence. The response to the peptides are given with a "+" indicating a positive response, and a "-" indicating a negative response.

This separation of these two different T cell activation assays was not simply a threshold dose phenomenon, and there is at least 6 orders of magnitude difference between the highest concentration of Asp-73 tested in the proliferation assay, and the lowest concentration that resulted in detectable IL-4 production. The IL-4 produced by stimulation with Asp-73 was able to provide B cell help; thus, B cell help can be achieved without the proliferation of T cells.

Potential Mechanisms

During the previous characterization of the Hb(64-76) determinant, we had identified the Glu-73 residue as being a T cell contact residue. Since, the Asp-73 peptide bound to I-E^k identically as the wild type peptide, a difference in the number of peptide/I-E^k complexes would not be an explanation since the number of Hb(64-76) or Asp-73 complexes would be the same. Th cells in addition to crosslinking of the T cell receptor, require a co-stimulatory signal. For Th2 cells, IL-1 can act as a co-stimulator. We next ascertained if IL-1 could synergize with the Asp-73 peptide to cause proliferation of the 2.102 T cells. The addition of IL-1 was able to restore the ability of the Asp-73 peptide to induce proliferation of the 2.102 cells. This result indicated that the Asp-73 peptide failed to induce a proliferative response because of a failure to induce the proper co-stimulatory signal.

The initial observation with the Asp-73 peptide was made using irradiated spleen cells as the antigen presenting cells(APC). It was possible that the two different peptides, wildtype and Asp-73 were being presented by different APC in the spleen cell preparation. We eliminated this possibility by testing the CD5$^+$ B cell line, CH27 for its ability to present the wild type and the Asp-73 peptides. The CH27 cells behaved identically to the spleen cells, in that both proliferation and IL-4 were produced by the wild type peptide, whereas only IL-4 was produced with the Asp-73 peptide.

Can the TCR differentially signal?

These results suggest that the T cell receptor is not simply an on/off switch, but that it may be able to deliver different signals. Our data would support a model of T cell activation that contained at least two steps. With the wild type peptide, the T cell receptor interacts with its ligand on the surface of an APC. Following the initial interaction, secondary adhesion/co-stimulatory molecules are induced on the T cell and the APC allowing both IL-4 production and T cell proliferation. When T cells are stimulated with the Asp-73 peptide, a lower affinity interaction between the TCR and its ligand causes a reduction in the secondary co-stimulator/adhesion molecules. The absence of these secondary interactions allows IL-4 production, but no proliferation. Two recent reports have shown that the TCR may signal through either the CD3-ε or the ξ chain[9,10]. While it is not known how these two different signalling pathways relate to the proliferation and lymphokine responses, the identification of multiple signalling pathways would allow differential activation of T cells.

References

1. R.G. Lorenz and P.M. Allen, Direct evidence for functional self protein/Ia-molecule complexes in vivo, *Proc.Natl.Acad.Sci.USA* 85:5220 (1988).

2. R.G. Lorenz and P.M. Allen, Thymic cortical epithelial cells lack full capacity for antigen presentation, *Nature* 340:557 (1989).

3. R.G. Lorenz and P.M. Allen, Thymic cortical epithelial cells can present self antigens in vivo, *Nature* 337:560 (1989).

4. B.D. Evavold, S.G. Williams, B.L. Hsu, S. Buus and P.M. Allen, Complete dissection of the Hb(64-76) determinant using Th1, Th2 clones, and T cell hybridomas, *J.Immunol.* 148:347 (1992).

5. B.S. Fox, C. Chen, E. Frage, C.A. French, B. Singh and R.H. Schwartz, Functionally distinct agretopic and epitopic sites. Analysis of the dominant T cell determinant of moth and pigeon cytochrome c with the use of synthetic peptide antigens, *J.Immunol.* 139:1578 (1987).

6. P.M. Allen, G.R. Matsueda, R.J. Evans, Jr. Dunbar,J.B., G.R. Marshall and E.R. Unanue, Dissection of a T cell antigenic epitope: Identification of the T cell and Ia contact residues, *Nature* 327:713 (1987).

7. R.G. Lorenz, A.N. Tyler and P.M. Allen, Reconstruction of the immunogenic peptide RNase(43-56) by identification and transfer of the critical residues into an unrelated peptide backbone, *J.Exp.Med.* 170:203 (1989).

8. B.D. Evavold and P.M. Allen, Separation of IL-4 production from Th cell proliferation by an altered T cell receptor ligand, *Science* 252:1308 (1991).

9. F. Letourneur and R.D. Klausner, Activation of T cells by a tyrosine kinase activation domain in the cytoplasmic tail of CD3 ζ, *Science* 255:79 (1992).

10. A.-M.K. Wegener, F. Letourneur, A. Hoeveler, T. Brocker, F. Luton and F. Malissen, The T cell receptor/CD3 complex is composed of at least two autonomous transduction modules, *Cell* 68:83 (1992).

CD28 RECEPTOR CROSSLINKING INDUCES

TYROSINE PHOSPHORYLATION OF PLC γ_1

Jeffrey A. Ledbetter and Peter S. Linsley

Bristol-Myers Squibb Pharmaceutical Research Institute
3005 First Avenue
Seattle, WA 98121

INTRODUCTION

The CD28 receptor has been widely recognized for its ability to regulate T cell responses to T cell receptor (TCR) stimulation (see [1] for review). CD28 crosslinking upregulates cytokine mRNA expression and secretion [2,3] when CD28 is co-ligated with the CD3/TCR and can directly induce IL2 Receptor α chain mRNA expression in resting T cells [4] in the absence of other signals. Despite the importance of the CD28 receptor, the mechanisms of CD28 signal transduction are not yet known. Here we provide evidence that phospholipase C (PLC) γ_1 is phosphorylated on tyrosine after CD28 ligation on primed T cells. In addition, the CD28-mediated induction of IL2 mRNA is inhibited by the tyrosine kinase inhibition herbimycin A, suggesting that the IL2 mRNA response may be partly caused by PLC γ_1 activation.

CD28 crosslinking on normal cells or on the Jurkat T cell leukemia results in an increase in cytoplasmic free calcium concentration [Ca2+]i [4-6]. This signal is mediated by release of calcium from internal stores and is associated with production of inositol phosphates, consistent with the activation of phosphoinositol-specific PLC [4]. However, PLC activation by CD28 ligation does not completely explain the CD28 biological effects, since (a) CD28 crosslinking on the cell surface is required for [Ca2+]i responses, but soluble 9.3 anti-CD28 mAb induces cytokine secretion in cells where no [Ca2+]i changes are detectable [7]; (b) CD28 stimulation increases cytokine secretion in T cells even when optimal concentrations of phorbol-12-myristate-13-acetate (PMA) and ionomycin are used [8]. Since these phamacologic compounds effectively mimmic the second messengers generated from PLC activation, the result indicates that CD28 receptor stimulates a response not simply restricted to PLC activation; and (c) CD28 co-stimulates IL2 mRNA expression in cells even in the presence of the inhibitor cyclosporine, whereas PMA and ionomycin induced mRNA expression is completely sensitive to cyclosporine [9].

Recent studies have indicated that CD28 crosslinking on T cells primed by overnight treatment with PMA induces tyrosine phosphorylation on proteins detectable by immunoblotting of whole cell lysates [10]. The priming is important for the response, since resting T cells were tested previously without detecting the signal [4]. These results

Mechanisms of Lymphocyte Activation and Immune Regulation IV: Cellular Communications
Edited by S. Gupta and T.A. Waldmann, Plenum Press, New York, 1992

23

suggest that CD28 may transmit information by activation of tyrosine kinase(s). Here we provide evidence that supports this hypothesis and show that PLC γ_1 is one of the tyrosine kinase substrates that is phosphorylated in response to CD28 ligation.

RESULTS

It has recently been realized by several groups that PLC γ_1 is activated in lymphocytes by tyrosine phosphorylation after stimulation of the TCR [11-14]. In fibroblasts, the enzymatic activity of PLC γ_1 is known to be directly regulated by tyrosine phosphorylation [15]. We therefore examined PLC γ_1 after CD28 crosslinking to see whether the same enzyme was involved in the CD28 receptor signal. For this experiment, T cells were primed by prior stimulation with phytohemagglutinin (PHA) and allowed to grow for 5-6 days. The PHA both enhances the CD28 signal and expands the population so that these experiments can be done using normal cells rather than cell lines.

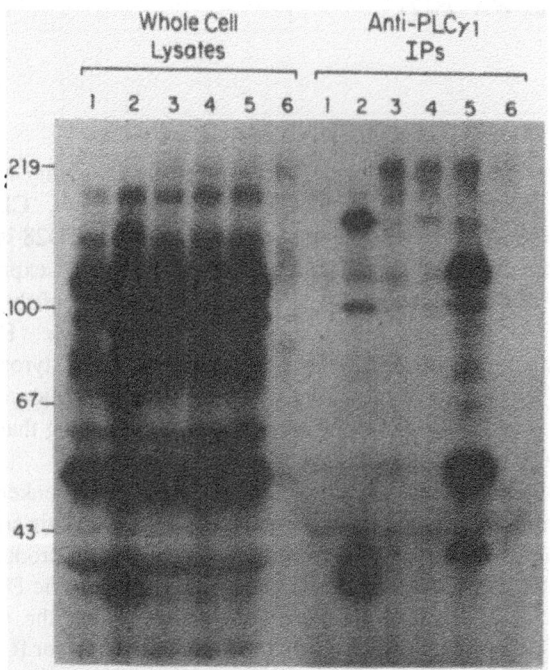

Figure 1. CD28-Induced Tyrosine Phosphorylation. Primed T cells were generated by expanding peripheral blood mononuclear cells with PHA-P (1 ug/ml, Wellcome Laboratories) for 6 days in RPM1 with 10% fcs. T cell blasts were washed and then stimulated at 4×10^7 cells/ml. Stimulations were (1) control (avidin only); (2) CD28 mAb 9.3-biotin 10 ug/ml for 5 min, followed by avidin (50 mg/ml) and sample preparation at 1 min; (3) co-culture with B7+CHO cells at a ratio of 1:40 (10^6 B7+CHO/4×10^7 T cells) for 5 min; (4) B7+CHO co-culture for 10 min; (5) B7+CHO co-culture for 15 min; (6) B7+CHO cells alone without T cells. Samples were prepared by lysis in 1% NP40, .25% DOC buffer as described [16]. Tyrosine phosphorylation was detected by immunoblotting with rabbit anti-p-tyr antibodies as previously described [17].

CD28 was crosslinked on the cell surface using biotin-conjugated mAb 9.3 followed by avidin, or by co-culture with Chinese hamster ovary (CHO) cells transfected with the B7 molecule, recently identified as a natural CD28 ligand [18,19]. The crosslinking with 9.3 induced tyrosine phosphorylation of multiple proteins detected in the whole cell lysates. PLC γ_1 and an associated 35/36 kDa protein were phosphorylated on tyrosine also and were detectable in the PLC γ_1 immunoprecipitation (Figure 1, lanes 1 and 2). Co-culture with B7+CHO cells also induced tyrosine phosphorylation detectable both in whole cell lysates and on PLC γ_1 IPs. The strong signal on PLC γ_1

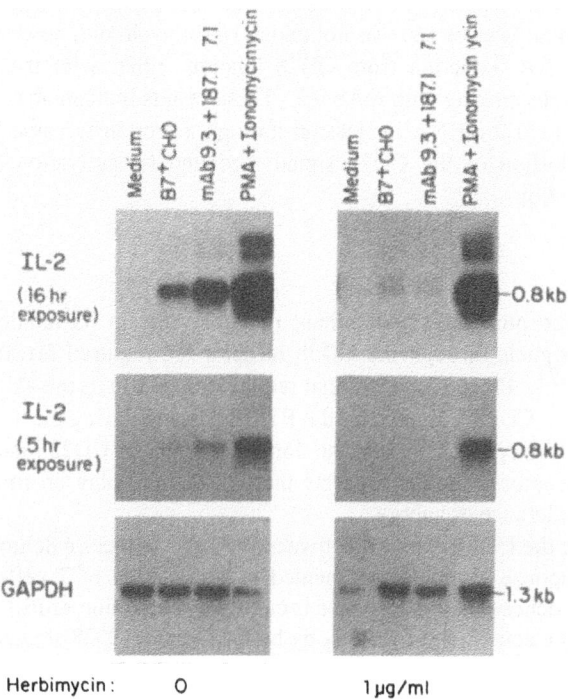

Figure 2. Induction of IL2 mRNA by CD28 ligation. Primed T cells were generated with PHA as described in the legend to figure 1. Blasts were washed and then incubated overnight with or without herbimycin A (1 mg/ml). Cells were then stimulated for 4 hours with B7+CHO cells, CD28 crosslinking with 9.3 plus 187.1, or with PMA (10 ng/ml) plus ionomycin (0.5 ug/ml). RNA was extracted as described [20] and blots for IL2 mRNA and GAPDH (control) mRNA were performed as described [18].

associated proteins seen after 15 min of co-culture (lane 5) may reflect amplification caused by additional tyrosine-phosphorylated proteins associating with PLC γ_1 in an activation complex. It is also possible that PLC γ_1 degredation occurred in this sample. The B7+CHO cells alone (lane 6) did not express tyrosine-phosphorylated PLC γ_1 and did not contribute significantly to the signal seen in whole cell lysates.

The results clearly show that the 9.3 mAb crosslinking induced a rapid response, but we have found the signal to be transient and not detectable after 10 min (data not shown). The signal from the B7$^+$CHO cells was slower and more prolonged. Figure 1 shows a response at for 15 min, and a clear signal in whole cell lysates has been seen in other experiments for up to 2 hours (data not shown).

One of the early responses after CD28 crosslinking is the expression of cytokine mRNAs. In the primed T cell PHA blasts, CD28 crosslinking induces IL2 mRNA expression without requiring additional TCR ligation, and the cells then proliferate in response to the IL2 signal (data not shown). We therefore examined the expression of IL2 mRNA in response to CD28 crosslinking either with B7$^+$CHO cells or with 9.3 + 187.1 (anti-K mAb). The cells were also stimulated with PMA + ionomycin for comparison, and the effect of the tyrosine kinase inhibitor herbimycin A on the response was measured.

The results show that herbimycin A has little or no inhibitory effect on the response to PMA plus ionomycin, indicating that herbimycin A was not inhibiting protein serine/threonine kinases and was not toxic. Herbimycin did, however, completely prevent the IL2 mRNA induction from CD28 ligation, either with the natural ligand (B7$^+$CHO) or with crosslinking with mAb 9.3. These results indicate that CD28 receptor signals are coupled to IL2 mRNA expression through a protein tyrosine kinase and are consistent with a portion of the CD28 signal mediated by activation of PLC γ_1 by tyrosine phosphorylation.

DISCUSSION

The CD28 receptor plays a dominant rgulatory role in its relationship to TCR receptor. That is, signals through the CD28 receptor are required for antigen induced responses of T cells [21]. B cell responses that require helper T cells are also dependent on CD28 sitmulation [22]. CD28 is ligated by the B7/BB1 ligand during the T cell interaction with antigen presenting cells [18,19], and the dominant effect of CD28 receptor signals in controlling TCR receptor response suggests that CD28 may play an important role in determining T cell tolerance or anergy.

Crosslinking the CD28 receptor activates PLC and induces calcium mobilization, and the calcium response is rapidly augmented by stimulation of T cells with PMA [4]. Here we provide evidence that this response is caused by activation of PLC γ_1 by tyrosine phosphorylation. The activation of PLC γ_1 by both TCR and CD28 suggests that PLC γ_1 is not the only mediator of the CD28 receptor signal. In additiona, modulation of CD3/TCR by soluble anti-CD3 prevents the subsequent CD28 receptor-induced calcium signal (unpublished data), suggesting that the CD28 coupling to PLC γ_1 activation requires CD3/TCR expression.

If PLC γ_1 is not the only mediator of the CD28 signal, what other pathways are involved? Although the precise answer is not yet known, our results support the conclusion that activation of a protein tyrosine kinase is required for the CD28 response. We detect a tyrosine phosphorylation signal from either B7$^+$CHO cells or from 9.3 crosslinking, and the tyrosine kinase inhibitor herbimycin A inhibits the CD28-induced mRNA expression without inhibiting the response to PMA + ionomycin. One approach to identifying additional CD28 response elements is to find tyrosine-phosphorylated substrates that are induced by CD28 ligation but not by anti-CD3 or anti-CD2 receptor ligation. Thus far, the substrates we have examined, including PLC γ_1, vav, and TCR zeta are phosphorylated on tyrosine by both TCR and CD28 receptor signalling (unpublished data). Another approach will be to identify the tyrosine kinases involved in the CD28 signal. In this regard, our results so far suggest that both p56^{lck} and p59^{fyn} play a role in CD28 signal transduction.

ACKNOWLEDGEMENTS

We thank Laura Grosmaire and Bill Brady for performing the stimulations and RNA blots, and Theta Tsu for help with the PLC γ_1 experiments.

REFERENCES

1. C.H. June, J.A. Ledbetter, P.S. Linsley, and C.B. Thompson, Role of the CD28 receptor in T cell activation, *Immunol. Today* 11:211 (1990).
2. T. Lindstein, C.H. June, J.A. Ledbetter, G. Stella, and C.B. Thompson, Regulation of lymphokine messenger RNA stability by a surface-mediated T cell activation pathway, *Science* 244:339 (1989).
3. C.B. Thompson, T. Lindsten, J.A. Ledbetter, S.L. Kunkel, H.A. Young, S.G. Emerson, J.M. Leiden, and C.H. June, CD28 activation pathway regulates the production of multiple T-cell-derived lymphokines/cytokines, *Proc. Natl. Acad. Sci. USA* 86:1333 (1989).
4. J.A. Ledbetter, J.B. Imboden, G.L. Schieven, L.S. Grosmaire, P.S. Rabinovitch, T. Lindsten, C.B. Thompson, and C.H. June, CD28 ligation in T cell activation: evidence for two signal transduction pathways, *Blood* 75:1531 (1990).
5. A. Weiss, B. Manger, and J. Imboden, Surgery between the T3/antigen receptor complex and Tp44 in the activation of human T cells, *J. Immunol.* 137:819 (1986).
6. J.A. Ledbetter, C.H. June, L.S. Grosmaire, and P.S. Rabinovitch, Crosslinking of surface antigens causes mobilization of intracellular ionized calcium in T lymphocytes, *Proc. Natl. Acad. Sci. USA* 84:1384 (1987).
7. J.A. Ledbetter, M. Parsons, P.J. Martin, J.A. Hansen, P.S. Rabinovitch, and C.H. June, Antibody binding to CD5 (Tp67) and Tp44 T cell surface molecules: effects on cyclic nucleotides, cytoplasmic free calcium, and cAMP-mediated suppression, *J. Immunol.* 137:3299 (1986).
8. C.H. June, J.A. Ledbetter, T. Lindsten, and C.B. Thompson, Evidence for the involvement of three distinct signals in the induction of IL-2 gene expression in human T lymphocytes, *J. Immunol.* 143:153 (1989).
9. C.H. June, J.A. Ledbetter, M.M. Gillespie, T. Lindsten, and C.B. Thompson, T-cell proliferation involving the CD28 pathway is associated with cyclosporine-resistant interleukin 2 gene expression, *Mol. Cell. Biol.* 7:4472 (1987).
10. P. Vandenberghe, G.J. Freeman, L.M. Nadler, M.C. Fletcher, M. Kamoun, L.A. Turka, J.A. Ledbetter, C.B. Thompson, and C.H. June, Antibody and B7/BB1-mediated ligation of the CD28 receptor induces tyrosine phosphorylation in human T cells, *J. Exp. Med.* (1992). (in press)
11. J.P. Secrist, L. Karnitz, and R.T. Abraham, T cell antigen receptor ligation induces tyrosine phosphorylation of phospholipase C-gamma1, *J. Biol. Chem.* 266:12135 (1991).
12. A. Weiss, G. Koretzky, R.C. Schatzman, and T. Kadlecek, Functional activation of the T cell antigen receptor induces tyrosine phosphorylation of phospholipase C-gamma1, *Proc. Natl. Acad. Sci. USA* 88:5484 (1991).
13. D.J. Park, H.W. Rho, and S.G. Rhee, CD3 stimulation causes phosphorylation of phospholipase C-gamma 1 on serine and tyrosine residues in a human T cell line, *Proc. Natl. Acad. Sci. USA* 88:5453 (1991).
14. C. Granja, L.-L. Lin, E.J. Yunis, V. Relias, and J.D. Dasgupta, PLCgamma1, a possible mediator of T cell receptor function, *J. Biol. Chem.* 266:16277 (1991).
15. S. Nishibe, M.I. Wahl, S.M.T. Hernandez-Sotomayor, N.K. Tonks, S.G. Rhee, and G. Carpenter, Increase of the catalytic activity of phospholipase C-gamma1 by tyrosine phosphorylation, *Science* 250:1253 (1990).
16. S.B. Kanner, A.B. Reynolds, and J.T. Parsons, Immunoaffinity purification of tyrosine-phosphorylated cellular proteins, *J. Immunol. Methods* 120:115 (1989).
17. J.A. Ledbetter, G.L. Schieven, F.M. Uckun, and J.B. Imboden, CD45 cross-linking regulates phospholipase C activation and tyrosine phosphorylation of specific substrates in CD3/Ti-stimulated T cells, *J. Immunol.* 146:1577 (1991).
18. P.S. Linsley, W. Brady, L. Grosmaire, A. Aruffo, N.K. Damle, and J.A. Ledbetter, Binding of the B cell activation antigen B7 to CD28 costimulates T cell proliferation and interleukin 2 mRNA accumulation, *J. Exp. Med.* 173:721 (1991).
19. P.S. Linsley, E.A. Clark, and J.A. Ledbetter, T cell antigen CD28 mediates adhesion with B cells by interacting with activation antigen B7/BB-1, *Proc. Natl. Acad. Sci. USA* 87:5031 (1990).
20. P. Chomczynski and N. Sacchi, Single step method of RNA isolation by acid guanidinium thiocyanate-phenol-chloroform extraction, *Anal. Biochem.* 162:156 (1987).
21. P.S. Linsley, W. Brady, M. Urnes, L.S. Grosmaire, N.K. Damle, and J.A. Ledbetter, CTLA-4 is a second receptor for the B cell activation antigen B7, *J. Exp. Med.* 174:561 (1991).
22. N.K. Damle, P.S. Linsley, and J.A. Ledbetter, Direct helper T cell-induced B cell differentiation involves interaction between T cell antigen CD28 and B cell activation antigen B7, *Eur. J. Immunol.* 21:1277 (1991).

STRUCTURE AND FUNCTION OF CD45:

A LEUKOCYTE-SPECIFIC PROTEIN TYROSINE PHOSPHATASE

Ian S. Trowbridge, Pauline Johnson,* Hanne Ostergaard,† and Nicholas Hole#

Department of Cancer Biology, The Salk Institute
P.O. Box 85800, San Diego, CA 92186-5800

INTRODUCTION

CD45 is a large abundant leukocyte-specific cell surface glycoprotein [1]. It is structurally heterogeneous consisting of a family of isoforms (Mr ~180-220K) that are distributed in characteristic cell-type specific patterns on different leukocyte subpopulations. Because of its leukocyte-specific tissue distribution, CD45 is a useful marker for the differential diagnosis of undifferentiated lymphoma [2]. The primary structures of rat, mouse and human CD45 deduced from cDNA sequencing in the mid-1980's revealed that the cytoplasmic domain of CD45 was unusually large[1], confirming earlier studies of a mutant CD45 molecule [3]. Charbonneau *et al.* showed in 1988 that two tandem imperfect repeats within the large cytoplasmic domain of CD45 had highly significant sequence similarity with PTP1B, a soluble ~37K human placental protein tyrosine phosphatase (PTP) [4]. Subsequently, CD45 was shown to have intrinsic PTP activity [5], suggesting that transmembrane PTPs might represent a novel class of receptors that oppose the action of protein tyrosine kinases (PTKs) and play a fundamental role in the regulation of cell growth and differentiation. Other putative transmembrane and intracellular PTPs in organisms as diverse as higher eukaryotes, *Drosophila*, yeast, bacteria, and viruses have now been identified and the complete primary structures of at least twenty PTPs have been determined (reviewed in Ref 6). CD45 is the best characterized of the transmembrane PTPs and serves as a prototype for this class of enzyme. An understanding of the function of CD45 in leukocyte physiology requires identification of its substrates, elucidation of the structural features of the molecule essential for PTP activity, and information about how its activity is regulated. This paper will review recent progress in

Present Addresses:
* Department of Microbiology, University of British Columbia,
  #300-6174 University Boulevard, Vancouver V6T 1W5,
  British Columbia, Canada
† Department of Immunobiology, 860 Medical Sciences Building,
  University of Alberta, Edmonton, Alberta T6G 2H7, Canada
AFRC Center for Animal Genome Research,
  University of Edinburgh, King's Buildings,
  West Mains Road, Edinburgh EH9 3JQ, U.K.

Mechanisms of Lymphocyte Activation and Immune Regulation IV: Cellular Communications
Edited by S. Gupta and T.A. Waldmann, Plenum Press, New York, 1992

29

defining the role of CD45 in lymphocyte signal transduction and summarize the current status of the rapidly growing PTP family.

ROLE OF CD45 IN LYMPHOCYTE SIGNAL TRANSDUCTION

Our current understanding of the role of CD45 in lymphocyte signal transduction rests heavily upon the use of CD45- mutant cell lines. Analysis of a series of CD45- mutant murine lymphoma cell lines derived by R. Hyman has provided compelling evidence that p56[lck], the src-related PTK found predominantly in T cells associated with either CD4 or CD8, is an *in vivo* substrate of CD45 [7]. The phosphotyrosine content of the cellular proteins of three independently-derived sets of CD45- mutant, CD45+ parental and CD45+ revertant murine lymphoma cell lines were compared by SDS-polyacrylamide gel electrophoresis of total cell lysates followed by immunoblotting with anti-phosphotyrosine antibodies. The rationale for this approach was that substrates of CD45 might be expected to have an increased phosphotyrosine content in cells lacking CD45 PTP activity assuming that the loss of CD45 was not compensated by concomitant changes in the activity of other PTPs or PTKs that act on the same substrates. Of the 3-5 major phosphotyrosine-containing proteins detected by immunoblotting, only one protein had significantly elevated levels of phosphotyrosine in the CD45- mutant cells. This protein was identified as p56[lck] by immunoprecipitation with a specific anti-p56[lck] rabbit antiserum. Interestingly, it was found that in all three cell lines, there was a selective increase in phosphorylation of Tyr-505, the negative regulatory site in p56[lck] equivalent to Tyr-527 in p60[src]. The implication of this observation is that CD45 influences the proliferative responses of T cells by dephosphorylation of p56[lck] on Tyr-505 thus increasing kinase activity and potentially providing a stimulatory growth signal. Consistent with this notion, functional analysis of CD45- mutants of CD4+ helper and CD8+ cytotoxic T cell clones shows that CD45 is required for the induction of a proliferative response following antigen stimulation [8,9]. It is also known that CD45 is necessary to trigger the initial metabolic events induced by antigen stimulation of both T and B cells [10-12]. Together, these data suggest that CD45 plays an essential positive role in signalling via the T cell receptor. It is of interest that a detectable phenotype is associated with the loss of CD45, as this indicates that other PTPs expressed in T cells, such as LRP and T cell PTP (see Fig 3), cannot fully compensate for the deficiency and suggests each of these PTPs will play a distinct role in lymphocyte physiology.

Independent evidence that p56[lck] is an *in vivo* substrate of CD45 was obtained from antibody-induced co-clustering of CD4 and CD45 [13]. The results of these experiments show that tyrosine phosphorylation of p56[lck] induced by crosslinking CD4 with anti-CD4 antibodies could be subsequently reversed by co-clustering with CD45 (Fig. 1). Important questions stemming from this work that are currently being actively investigated include whether the phosphotyrosine content of other members of the src family of PTKs are elevated in CD45- mutant lymphomas and whether the phosphotyrosine level of p56[lck] is increased in the CD45- mutant helper and cytotoxic murine cell lines subjected to functional analysis. The answers to these questions will help in evaluating the physiological significance of the dephosphorylation of Tyr-505 of p56[lck] by CD45.

REGULATION OF CD45 ACTIVITY

The structure of CD45 is consistent with that of a receptor analogous to a growth factor receptor PTK. This implies that the activity of CD45, and possibly other transmembrane PTPs, might be regulated by interaction of its external domain with a specific ligand. To address this issue, we have expressed the external domain of the B cell isoform of murine CD45 as a soluble secreted protein in Chinese hamster ovary cells using a glutamine synthetase expression system [14]. The recombinant external domain can be produced in high yield (5-20mg/l) and can be purified 14-fold to homogeneity with a yield of ~30% by pea lectin affinity chromatography followed by fractionation on a Superose 6 FPLC column. The 160K recombinant protein is heavily glycosylated (carbohydrate accounting for ~ 65% of its M_r), and has the same immunoreactivity as the external domain of naturally-occurring CD45. Interestingly, the ~69K nonglycosylated external domain produced by cells treated with

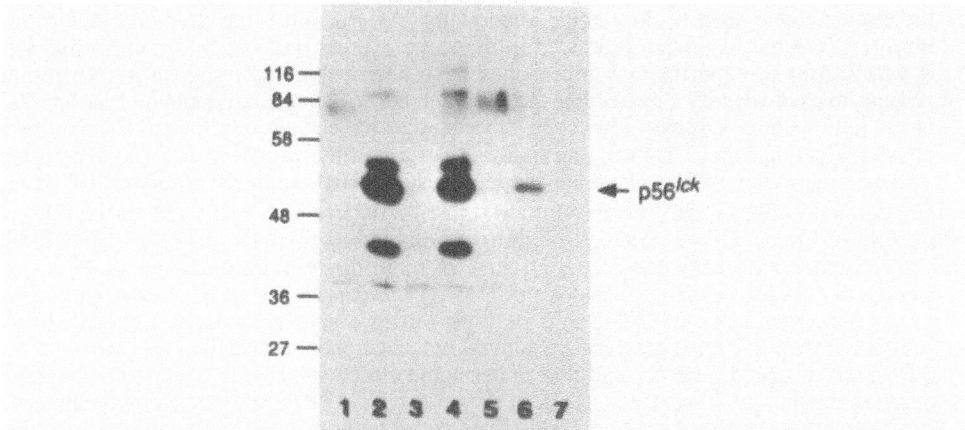

Fig. 1. Reversal of increased tyrosine phosphorylation of p56^{lck} induced by CD4 clustering by coclustering with anti-CD45 MAb. AODH 7.1 cells were crosslinked with anti-CD4 MAb (GK 1.5) and anti-Ig to induce tyrosine phosphorylation of p56^{lck}. After induction for 2 min, anti-CD45 MAb (I3/2) was added, and the cells were incubated for an additional 13 min. Controls of anti-CD4 alone, anti-CD45 alone, and anti-CD4 with anti-CD45 together were also done. (Lane 1) Second antibody alone for 15 min; (lane 2) anti-CD4 for 2 min; (lane 3) anti-CD4 with anti-CD45 for 2 min; (lane 4) anti-CD4 for 15 min; (lane 5) anti-CD4 with anti-CD45 for 15 min; (lane 6) anti-CD4 for 2 min, then addition of anti-CD45 with a further 13-min incubation; (lane 7) anti-CD45 for 15 min. Reproduced with permission from Ref. 13.

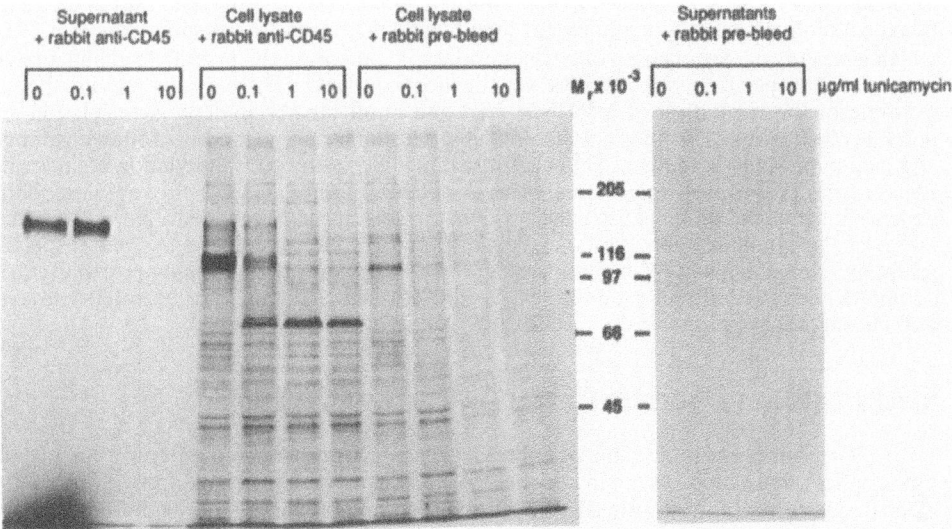

Fig. 2. Glycosylation of soluble secreted recombinant murine CD45R(ABC) external domain produced in CHO cells. CHO cells expressing murine CD45R(ABC) external domain were selected as described in Ref. 14. Cells were metabolically labelled with [^{35}S] methionine in the presence or absence of varying concentrations of tunicamycin, and the supernatants and cell lysates subjected to immunoprecipitation with either rabbit polyclonal antibody to murine CD45R(ABC) external domain or normal rabbit serum. Immunoprecipitates were run under reducing conditions on an 8% SDS-PAGE gel.

tunicamycin is not efficiently secreted but can be immunoprecipitated from cell lysates (Fig. 2). The B cell isoform of the external domain has a Stoke's radius of 80±2 A and a frictional coefficient of 2.1±0.1, consistent with the rod-like structure of the external domain of rat thymocyte CD45 seen by low-angle shadowing [15]. Recently, the thymocyte isoform of murine CD45 has also been produced in the same expression system in comparable yield. Several different experimental approaches have been used employing these recombinant proteins to try to identify a ligand that specifically binds to the external domain of murine CD45 but to date without success. However, other studies have suggested that CD22 may be the physiological ligand of CD45R0, the smallest isoform of human CD45 [16]. The experimental evidence supporting this conclusion is indirect, resting mainly upon the ability of UCHL-1, a monoclonal antibody that reacts with an antigenic determinant restricted to the CD45R0 isoform of human CD45, to block binding of soluble recombinant CD22 immunoglobulin hybrid protein to T cells [16]. In the absence of more direct evidence, such as the specific binding of CD22 to cells transfected with CD45R0, this data should be interpreted cautiously.

The abundance of CD45 and its high intrinsic specific activity against artificial substrates *in vitro* (20) suggests that its activity must be tightly regulated *in vivo* whether or not this involves ligand binding, and that under some circumstances, is likely to be subject to negative regulation. Recently, it has been shown that CD45 PTP activity is downregulated by treatment of a variety of murine T cell lines and thymocytes with the calcium ionophore, ionomycin [17]. Treatment of mouse thymocytes with 1-2 μM ionomycin inhibits CD45 activity by >90% and similar results are obtained with functional T cell lines. The effect of ionomycin on spleen cells, however, is much less marked. The reason for this result is not known but may reflect the fact that only a small subpopulation of spleen cells are sensitive to ionomycin treatment.

When cells are treated with ionomycin, there is an almost instantaneous influx of Ca^{2+}; however, the decrease in CD45 activity occurs much more slowly and is only maximal after 30-40 min. The slow kinetics of the ionomycin effect suggests that the decrease in CD45 activity is not directly due to Ca^{2+} but, instead, is the result of a cascade of secondary events triggered by the increase in intracellular Ca^{2+}. Because the activity of PTKs was known to be regulated by phosphorylation, the possibility that changes in phosphorylation of CD45 may be involved in the loss of PTP activity induced by ionomycin treatment was investigated. It was found that treatment of thymocytes with ionomycin significantly decreased the phosphorylation of CD45 which is heavily phosphorylated on serine residues [18]. Interestingly, peptide mapping analysis suggested that specific serine residues were dephosphorylated. These results provide the first example of the regulation of a PTP *in vivo* and implicate serine dephosphorylation as a possible mechanism. The transient increase in intracellular Ca^{2+} that occurs following antigen stimulation may downregulate CD45 activity allowing tyrosine phosphorylation of substrates required to drive a proliferative response to occur. Further, the Ca^{2+} - and calmodulin-dependent serine/threonine phosphatase calcineurin has recently been identified as a common target of the immunosuppressants, cyclosporin and FK506 complexed to their respective receptors, cyclophilin and FK binding protein [19-21]. This raises the intriguing possibility these compounds might indirectly inhibit CD45 activity and that this may be related to their immunosuppressive properties.

STRUCTURAL REQUIREMENTS FOR CD45 ACTIVITY

The cytoplasmic domain of murine CD45 has been expressed using an in vitro transcription/ translation system in which RNA transcripts were generated from murine CD45 cDNA constructs in pBluescriptSK, using the T3 promoter and T3 polymerase [22]. CD45 mRNA was then translated in a rabbit reticulocyte lysate system in the presence of [^{35}S]methionine to produce biosynthetically-labeled recombinant CD45 protein. The *in vitro* translated proteins were purified by immunoprecipitation with a rabbit antiserum raised by immunization with murine CD45 cytoplasmic domain produced in a baculovirus expression system [7]. As the recombinant proteins were metabolically-labeled and purified by immunoprecipitation, the amount of translated CD45 protein in each assay could be quantitated, allowing the relative PTP activities of wildtype and mutant proteins to be calculated. The limit of detection of mutant CD45 proteins was ~4% that of wildtype activity.

It was found that CD45 PTP domain I was not active without the presence of an intact PTP domain II. Thus, CD45 differs from LAR and HPTP-α as the first domains of these related transmembrane PTPs are active when expressed alone [23-25]. It is unclear whether the second PTP domain of CD45 is required for proper folding of the first domain, or whether it acts as a positive regulatory element. As previously reported [23], the second PTP domain of CD45 did not have detectable PTP activity *in vitro*. It is known that there is a highly conserved cysteine residue within the carboxy-terminal region of active PTP domains which is essential for activity and is thought to form a thiol-phosphate intermediate during catalysis [26, 27]. This essential cysteine residue is surrounded by a region of 10 residues which is the most highly conserved feature of PTP domains. The consensus sequence for this conserved segment is [I/V]HCSAGVGR[S/T]G. Point mutations within this region of the first PTP domain of CD45 either abolish activity or decrease it significantly and it is likely that this region forms part of the active site. The second subdomain of CD45 is unusual in that five of the residues within this region differ from the consensus sequence. Single substitutions of three of the conserved residues in this region of the first PTP domain of CD45 with the nonconserved residues found at the equivalent positions in the second domain were found to abolish activity explaining why the second domain is inactive against substrates for the first domain. Modifying this region of the second PTP domain to conform to the consensus sequence, however, is not sufficient to generate activity indicating the second PTP domain of CD45 lacks other structural features required for activity.

OTHER MEMBERS OF THE PTP FAMILY

The PTP family continues to grow rapidly in size and diversity and can be subdivided into transmembrane and intracellular PTPs. Currently, the complete primary structures of at least ten putative transmembrane PTPases and ten intracellular PTPs have been deduced from cDNA nucleotide sequences (Fig. 3). With the exception of HPTP-β and DPTP 10D, the cytoplasmic domains of the transmembrane PTPs are similar in structure and have tandem catalytic domains. The cytoplasmic domains of LAR, CD45, HPTP-α and HPTP-β have been expressed as enzymatically active recombinant proteins [7,22-25]. The significance of the tandem catalytic domains of the transmembrane PTPases is unclear. The second catalytic domains of LAR and CD45 are not active against the artificial substrates commonly used to assay PTP activity *in vitro*, and the second domain of HPTP-γ and rat PTP18 [28,29] and a related *Drosophila* PTP that has several isoforms [30-32] lack the essential cysteine residue. The external domain of CD45 is unique, whereas those of LAR, DLAR, DPTP, DPTP 99A and HPTP-β are related, being composed of repeating units of immunoglobulin and fibronectin type III repeats (Fig. 3). This has led to the suggestion that this latter group of PTPs are involved in cell-cell interactions [28]. Until recently, CD45 was the only receptor-like PTPase known to have a highly restricted tissue distribution, but now there are three reports of transmembrane PTPs that are selectively expressed in the central nervous system neurons of *Drosophila*; [30-32]. LAR and LRP (HPTP-α), appear to be expressed on a wide variety of tissues based on Northern blot analysis of mRNA levels [33-36] and the tissue distributions of the other mammalian receptor-like PTPs have not been reported. Antibodies specific for the external domains of each of the transmembrane PTPs will be required for further analysis of their tissue distribution and function.

Seven intracellular mammalian PTPs have been identified; PTP 1B and T cell PTP have a broad tissue distribution and contain ~430 amino acids with a single PTPase domain located in their amino-terminal region [4, 37-41]. The carboxy-terminal regions of these two PTPs appear to be important in determining their intracellular localization and regulation of their enzymic activity. The full-length intracellular PTPs are associated with the particulate fraction of cell homogenates and have hydrophobic carboxy-termini that may serve as membrane anchors. A novel cDNA encoding STEP, a putative brain-specific intracellular PTP of ~369 amino acids, was isolated from a rat striatal cDNA library [42]. PEG [43] isolated from a megakaryocyte cDNA library and PTPH1 [44] are closely related intracellular PTPs of 926 and 913 amino acids respectively that both have a region of similarity in their amino-terminal segments with the amino-terminal regions of the cytoskeletal-associated proteins band 4.1, ezrin and talin. Another human intracellular PTP, designated PTP1C, has been identified that, interestingly, has in its amino-terminal region two SH2 domains [45]. Recently, it has been shown that cdc25

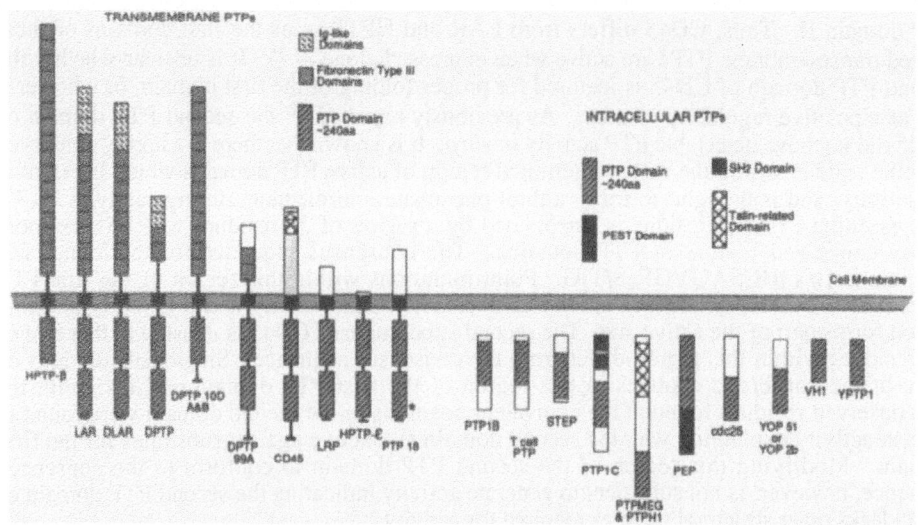

Fig. 3. Schematic representation of the PTP family. PTPs are oriented with their NH^2 termini toward the top of the diagram. Transmembrane PTPs are type I membrane proteins with an amino-terminal extracellular domain (shown as extending above the cell membrane), a single transmembrane region, and a carboxyl-terminal cytoplasmic domain. HPTP-α, -β, and -ε, LAR, CD45, PTP 1B, T cell PTP, STEP, PTP 1C, PTP MEG, and PTP H1 have been identified in the human and, in some cases, other mammalian species. PEP in the mouse, PTP 18 in the rat, DLAR, DPTP, DPTP 10D, and DPTP 99A in *Drosophila*, YOP 51 and YOP 26 in *Yersinia*, VH1 in vaccinia virus, and YPTP in yeast. LRP (34) has been cloned by several groups, and its various synonyms are HLPR (35), HPTP-α (36), and RPTP-α (37). The *asterisk* by PTP 18 indicates that an aspartate residue has replaced the conserved cysteine residue found in the carboxyl-terminal region of the PTP II subdomains of other transmembrane PTPs. See text for further details.

protein, which is known to control entry into mitosis by triggering tyrosine dephosphorylation of the cdc2 protein kinase, can directly dephosphorylate p34^{cdc2} *in vitro* [46-49].

Two other novel PTPs have been identified in pathogenic organisms (reviewed in Ref. 50). One was identified by a computer search for sequences related to conserved PTP sequences in the bacterial genus *Yersinia*, which is comprised of three species of bacteria that are causative agents in human disease. Each *Yersinia* species contains a plasmid encoding proteins associated with virulence. In the case of *Y. pseudotuberculosis*, it has been shown that the Yop H gene on such a plasmid encodes a PTP that is an essential virulence determinant and that can dephosphorylate tyrosine residues of proteins in host macrophages. The other PTP was identified in vaccinia virus, and interestingly, although this protein, VH1, is clearly a member of the PTP family and can dephosphorylate tyrosine residues, it also has significant activity against phosphoserine. An intracellular PTP lacking a carboxy-terminal regulatory region has been identified in the budding yeast *S. cerevisiae* [51]. Another putative 555 amino acid PTP has also been identified in the fission yeast, *S. pombe*, but the recombinant protein expressed in bacteria lacked detectable enzymatic activity [52].

CONCLUDING REMARKS

An essential role for CD45 in lymphocyte proliferative responses has been clearly established. However, we are only at the beginning of defining the biological role of individual PTPs and understanding how their activity is regulated. A key question remains whether CD45 and other transmembrane PTPs are receptors. In addition to ligand binding, the activity of PTPs may be regulated by other mechanisms. Whether some PTPs function as anti-oncogenes by opposing the action of transforming PTKs is another important issue under active investigation. The identification of physiological substrates for each of the PTPs will provide insight into the role of PTPs in signal transduction. In the next few years, it seems likely data that PTPs will be shown to be as important as their counterparts, PTKs, in the regulation of cell growth and metabolism.

ACKNOWLEDGEMENTS:

This work was supported by Grant CA 17733 from the National Cancer Institute.

REFERENCES

1. M.L. Thomas, The leukocyte common antigen family, *Ann. Rev. Immunol.* 7:339 (1989).
2. H. Battifora and I.S. Trowbridge, A monoclonal antibody useful for the differential diagnosis between malignant lymphoma and nonhematopoietic neoplasms, *Cancer* 51:816 (1983).
3. R. Hyman, I. Trowbridge, V. Stallings, and J. Trotter, Revertant expressing a structural variant of T200 glycoprotein, *Immunogenetics* 15:413 (1982).
4. H. Charbonneau, N.K. Tonks, K.A. Walsh, and E.H. Fischer, The leukocyte common antigen (CD45): a putative receptor-linked protein tyrosine phosphatase, *Proc. Natl. Acad. Sci. USA* 85:7182 (1988).
5. N.K. Tonks, H. Charbonneau, C.D. Diltz, E.H. Fischer, and K.A. Walsh, Demonstration that the leukocyte common antigen CD45 is a protein tyrosine phosphatase, *Biochemistry* 27:8696 (1988).
6. I.S. Trowbridge, CD45: a prototype for transmembrane protein tyrosine phosphatases, *J. Biol. Chem.* 266:23517 (1991).
7. H.L. Ostergaard, D.A. Shackelford, T.R. Hurley, P. Johnson, R. Hyman, B.M. Sefton, and I.S. Trowbridge, Expression of CD45 alters phosphorylation of the *lck*-encoded tyrosine protein kinase in murine lymphoma T-cell lines, *Proc. Natl. Acad. Sci. USA* 86:8959 (1989).
8. J.T. Pingel and M.L. Thomas, Evidence that the leukocyte-common antigen is required for antigen-induced T lymphocyte proliferation, *Cell* 58:1055 (1989).
9. C.T. Weaver, J.T. Pingel, J.O. Nelson, and M.L. Thomas, CD8+ T cell clones deficient in the expression of the CD45 protein tyrosine phosphatase have impaired responses to T cell receptor stimuli, *Mol. Cell. Biol.* 11:4415 (1991).
10. G.A. Koretzky, J. Picus, M.L. Thomas, and A. Weiss, Tyrosine phosphatase CD45 is essential for coupling T-cell antigen receptor to the phosphatidyl inositol pathway, *Nature* 346:66 (1990).
11. G.A. Koretzky, J. Picus, T. Schultz, and A. Weiss, Tyrosine phosphatase CD45 is required for T-cell antigen receptor and CD2-mediated activation of a protein tyrosine kinase and interleukin 2 production, *Proc. Natl. Acad. Sci. USA* 88:2037 (1991).
12. L.B. Justement, K.S. Campbell, N.C. Chien, and J.C. Cambier, Regulation of B cell antigen receptor signal transduction and phosphorylation by CD45, *Science* 252:1839 (1991).
13. H.L. Ostergaard and I.S. Trowbridge, Coclustering CD45 with CD4 or CD8 alters the phosphorylation and kinase activity of p56^{lck}, *J. Exp. Med.* 172:347 (1990).
14. I.S. Trowbridge, H. Ostergaard, D. Shackelford, N. Hole, and P. Johnson, CD45: a Leukocyte-specific protein tyrosine phosphatase, *Advances in Prot. Phosphatases* 6:227 (1991).
15. G.R. Woollett, A.F. Williams, and D.M. Shotton, Visualization by low-angle shadowing of the leucocyte-common antigen. A major cell surface glycoprotein of lymphocytes, *EMBO J.* 4:2827 (1985).
16. I. Stamenkovic, D. Sgroi, A. Aruffo, M.S. Sy, and T. Anderson, The B lymphocyte adhesion molecule CD22 interacts with leukocyte common antigen CD45RO on T cells and α2-6 sialyltransferase, CD75, on B cells, *Cell* 66:1133 (1991).
17. H.L. Ostergaard and I.S. Trowbridge, Negative regulation of CD45 protein tyrosine phosphatase activity by ionomycin in T cells, *Science* 253:1423 (1991).

18. M.B. Omary and I.S. Trowbridge, Disposition of T200 glycoprotein in the plasma membrane of a murine lymphoma cell line, *J. Biol. Chem.* 255:1662 (1980).

19. J. Friedman and I. Weissman, Two cytoplasmic candidates for immunophilin action are revealed by affinity for a new cyclophilin: one in the presence and one in the absence of CsA, *Cell* 56:799 (1991).

20. J. Liu, J.D. Farmer, Jr., W.S. Lane, J. Friedman, I. Weissman, and S.L. Schreiber, Calcineurin is a common target of cyclophilin-cyclosporin A and FKBP-FK506 complexes, *Cell* 66:807 (1991).

21. F. McKeon, When worlds collide: immunosuppressants meet protein phosphatases, *Cell* 66:823 (1991).

22. P. Johnson, H.L. Ostergaard, C. Wasden, and I.S. Trowbridge, Mutational analysis of CD45, *J. Biol. Chem.* 267, in press (1992).

23. M. Streuli, N.X. Krueger, T. Thai, M. Tang, and H. Saito, Distinct functional roles of the two intracellular phosphatase like domains of the receptor-linked protein tyrosine phosphatases LCA and LAR, *EMBO J.* 9:2399 (1990).

24. H. Cho, S.E. Ramer, M. Itoh, D.G. Winkler, E. Kitas, W. Bannwarth, P. Burn, H. Saito, and C.T. Walsh, Purification and characterization of a soluble catalytic fragment of the human transmembrane leukocyte antigen related (LAR) protein tyrosine phosphatase from an *Escherichia coli* expression system, *Biochemistry* 30:6210 (1991).

25. Y. Wang and C.J. Pallen, The receptor-like protein tyrosine phosphatase HPTPα has two active catalytic domains with distinct substrate specificities, *EMBO J.* 11:3231 (1991).

26. M. Streuli, N.X. Krueger, A.Y.M. Tsai, and H. Saito, A family of receptor-linked protein tyrosine phosphatases in humans and *Drosophila*, *Proc. Natl. Acad. Sci. USA* 86:8698 (1989).

27. K.L. Guan and J.E. Dixon, Protein tyrosine phosphatase catalysis proceeds via a cysteine-phosphate intermediate, *J. Biol. Chem.* 266:17026 (1991).

28. N.X. Krueger, M. Streuli, and H. Saito, Structural diversity and evolution of human receptor-like protein tyrosine phosphatases, *EMBO J.* 9:3241 (1990).

29. K. Guan and J.E. Dixon, Protein tyrosine phosphatase activity of an essential virulence determinant in *Yersinia*, *Science* 249:553 (1990).

30. X. Yang, K.T. Seow, S.M. Bahri, S. H. Oon, and W. Chia, Two drosophila receptor-like tyrosine phosphatase genes are expressed in a subset of developing axons and pioneer neurons in the embryonic CNS, *Cell* 67:661 (1991).

31. S.-S. Tian, P. Tsoulfas, and K. Zinn, Three receptor-linked protein-tyrosine phosphatases are selectively expressed on central nervous system axons in the *Drosophila* embryo, *Cell* 67:675 (1991).

32. I.K. Hariharan, P.-T. Chuang, and G.M. Rubin, Cloning and characterization of a receptor-class phosphotyrosine phosphatase gene expressed on central nervous system axons in *Drosophila melanogaster*, *Proc. Natl. Acad. Sci. USA* 88:11266 (1991).

33. M. Streuli, N.X. Krueger, L.R. Hall, S.F. Schlossman, and H. Saito, A new member of the immunoglobulin superfamily that has a cytoplasmic region homologous to the leukocyte common antigen, *J. Exp. Med.* 168:1553 (1988).

34. R.J. Matthews, E.D. Cahir, and M.L. Thomas, Identification of an additional member of the protein-tyrosine-phosphatase family: Evidence for alternative splicing in the tyrosine phosphatase domain, *Proc. Natl. Acad. Sci. USA* 87:4444 (1990).

35. F.R. Jirik, N.M. Janzen, I.G. Melhado, and K.W. Harder, Cloning and chromosomal assignment of a widely expressed human receptor-like protein-tyrosine phosphatase, *FEBS* 273:239 (1990).

36. J. Sap, P.D. D'Eustachio, D. Givol, and J. Schlessinger, Cloning and expression of a widely expressed receptor tyrosine phosphatase, *Proc. Natl. Acad. Sci. USA* 87:6112 (1990).

37. K. Guan, R.S. Huan, S.J. Watson, R.L. Geahlen, and J.E. Dixon, Cloning and expression of a protein-tyrosine-phosphatase, *Proc. Natl. Acad. Sci. USA* 87:1501 (1990).

38. J. Chernoff, A.R. Schievella, C.A. Jost, R.L. Erikson, and B.G. Neel, Cloning of a cDNA for a major human protein-tyrosine-phosphatase, *Proc. Natl. Acad. Sci. USA* 87:2735-2739 (1990).

39. D.E. Cool, N.K. Tonks, H. Charbonneau, K.A. Walsh, E.H. Fischer, and E.G. Krebs, cDNA isolated from a human T-cell library encodes a member of the protein-tyrosine-phosphatase family, *Proc. Natl. Acad. Sci. USA* 86:5257 (1989).

40. G. Swarup, S. Kamatkar, V. Radha, and V. Rema, Molecular cloning and expression of a protein-tyrosine phosphatase showing homology with transcription factors Fos and Jun, *FEBS* 280:65 (1991).

41. B. Mosinger, Jr., U. Tillmann, H. Westphal, and M.L. Tremblay, Cloning and characterization of a mouse cDNA encoding a cytoplasmic protein-tyrosine-phosphatase, *Proc. Natl. Acad. Sci. USA* 89:499 (1992).

42. P.J. Lombroso, G. Murdoch, and M. Lerner, Molecular characterization of a protein-tyrosine-phosphatase enriched in striatum, *Proc. Natl. Acad. Sci. USA* 88:7242 (1991).

43. M. Gu, J.D. York, I. Warshawsky, and P.W. Majerus, Identification, cloning, and expression of a cytosolic megakaryocyte protein-tyrosine-phosphatase with sequence homology to cytoskeletal protein 4.1, *Proc. Natl. Acad. Sci. USA* 88:5867 (1991).

44. Q. Yang and N.K. Tonks, Isolation of a cDNA clone encoding a human protein-tyrosine phosphatase with homology to the cytoskeletal-associated proteins band 4.1, ezrin, and talin, *Proc. Natl. Acad. Sci. USA* 88:5949 (1991).

45. S.H. Shen, L. Bastien, B.I. Posner, and P. Chretien, A protein-tyrosine phosphatase with sequence similarity to the SH2 domain of the protein-tyrosine kinases, *Nature* 352:736 (1991).

46. U. Strausfeld, J.C. Labbé, D. Fesquet, J.C. Cavadore, A. Picard, K. Sadhu, P. Russell, and M. Dorée, Dephosphorylation and activation of a p34^{cdc2} cyclin-B-complex *in vitro* by human cdc25 protein, *Nature* 351:242 (1991).

47. W.G. Dunphy and A. Kumagai, The cdc25 protein contains an intrinsic phosphatase activity, *Cell* 67:189 (1991).

48. J. Gautier, M.J. Solomon, R.N. Booker, J.F. Bazan, and M.W. Kirschner, Cdc25 is a specific tyrosine phosphatase that directly activates p34^{cdc2}, *Cell* 67:197 (1991).

49. J.B.A. Millar, C.H. McGowan, G. Lenaers, R. Jones, and J. Russell, p80^{cdc25} mitotic inducer is the tyrosine phosphatase that activates p34^{cdc2} kinase in fission yeast, *EMBO J.* 10:4301 (1991).

50. J.C. Clemens, K. Guan, J.B. Bliska, S. Falkow, and J.E. Dixon, Microbial pathogenesis and tyrosine dephosphorylation: Surprising bedfellows, *Mol. Microbiol.*, in press (1992).

51. K. Guan, R.J. Deschenes, H. Qin, and J.E. Dixon, Cloning and expression of a yeast protein tyrosine phosphatase, *J. Biol. Chem.* 266:12964 (1991).

52. S. Ottilie, J. Chernoff, G. Hannig, C.S. Hoffman, and R.L. Erikson, A fission-yeast gene encoding aprotein with features of protein-tyrosine-phosphatases *Proc. Natl. Acad. Sci. USA* 88:3455 (1991).

MULTIDRUG RESISTANT GENE 1 PRODUCT IN HUMAN T CELL SUBSETS: ROLE OF PROTEIN KINASE C ISOFORMS AND REGULATION BY CYCLOSPORIN A

Sudhir Gupta, Takashi Tsuruo[1] and Sastry Gollapudi

Division of Basic and Clinical Immunology
University of California
Irvine, CA 92717
[1]Cancer Chemotherapy Center, Japanese Foundation for
Cancer Research, Tokyo 170

INTRODUCTION

The *mdr 1* gene codes for a 170 kd P-glycoprotein (Pgp). The Pgp is a plasma membrane glycoprotein that shares extensive homology with numerous bacterial and eukaryotic transport proteins.[1-3] The sequence encoding Pgp reveals a tandemly duplicated molecule of approximately 1280 amino acids. Each half consists of a large hydrophobic domain containing 6 putative membrane spanning α-helices and a highly conserved hydrophilic domain with an ATP binding site.[4] P-glycoprotein has been implicated to play a major role in acquired multidrug resistance in cancer cells,[5,6] in chloroquine-resistant *Plasmodium falciparum* infection,[7,8] and possibly in human immunodeficiency virus 1 resistance to azidothymidine.[9]

Recently developed specific antibodies against Pgp and cDNA probes have been utilized to analyze the distribution of Pgp in normal human and rodent tissues (reviewed in 6). Although the physiologic role of Pgp is unclear, its apical and polar distribution in small and large intestine, renal proximal tubular cells, bile and pancreatic ductule cells, etc., suggests a secretory function of Pgp and perhaps its role to clear the cells of xenobiotics.

In the present paper we will review the data regarding expression and function of Pgp in human T cell subsets and present data regarding the regulation of Pgp by cyclosporin A (CsA) and involvement of protein kinase C isoforms in the expression of Pgp.

RESULTS

Expression of Pgp in Human T Cell Subsets

Dual color flow cytometric analysis of freshly isolated and PHA-activated peripheral blood mononuclear cells (MNC), using anti-CD3, anti-CD4 or anti-CD8 monoclonal antibodies and MRK-16 monoclonal antibody against Pgp,[10] revealed that CD8+ T cells preferentially express Pgp.[11] Approximately 80-90% CD8+ T cells and approximately 25% of CD4+ T cells expressed Pgp. Increased cell surface expression of Pgp, following PHA-activation, was associated with an increase in *mdr 1* mRNA as demonstrated by polymerase chain reaction[11] and Northern blot analysis (Figure 1).

To determine whether the expression of Pgp is cell-cycle dependent, MNC were activated with PHA in the presence or absence of hydroxyurea. Cell cycle analysis and Pgp expression were analyzed with FACScan. Table 1 shows that hydroxyurea (HU) blocked the transition of

Mechanisms of Lymphocyte Activation and Immune Regulation IV: Cellular Communications
Edited by S. Gupta and T.A. Waldmann, Plenum Press, New York, 1992

39

MNC from G1 to S phase of cell cycle; however, no effect was observed on Pgp expression. These data suggest that most of the membrane Pgp augmentation following T cell activation occurs in G1-S phase of cell cycle. A definitive study, however will require the analysis of cells sorted on the basis of cell cycle phase.

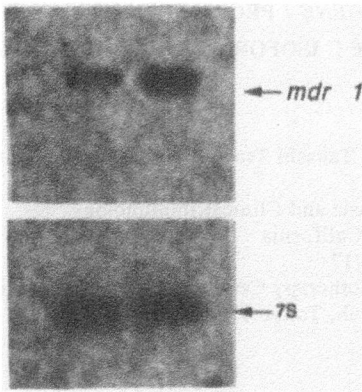

Figure 1. Northern blot analysis of *mdr 1* mRNA in MNC stimulated with PHA for 48 hours

Table 1. Relationship between P-glycoprotein expression and cell cycle phase

Experimental Conditions	Cell Cycle Phases			Pgp Positive Cells (%)
	Go	G1-S	G2 + M	
MNC	95.9	1.5	2.6	0.8
MNC + PHA	76.5	16.1	7.4	23.0
MNC + PHA + HU	91.3	3.9	4.8	21.0

To determine whether any transcriptional and/or translational events are involved in PHA-induced augmentation of Pgp, MNC were incubated in the presence or absence of cycloheximide and actinomycin D for 30 minutes, washed and then stimulated for 48 hours with PHA. No cell death was observed in any of the samples. P-glycoprotein expression in T cells was analyzed by dual color flow cytometry, using anti-CD3 and MRK-16 monoclonal antibodies. Table 2 shows that both actinomycin D and cycloheximide inhibited PHA-induced augmentation of membrane Pgp. These observations indirectly suggest that both transcriptional and translational events are involved in PHA-induced augmentation of Pgp. Nuclear run-off assay is required to precisely define the transcriptional and translational events. Furthermore, *in situ* hybridization studies are needed to determine whether PHA activation also leads to increase gene copies of *mdr 1*.

Protein Kinase C is Involved in an Augmentation of Pgp

Protein kinase C (PKC) is involved in signal transduction during T cell activation.[12] To determine whether PKC is involved in PHA-induced augmentation of Pgp, three approaches were used. First, MNC were activated with low concentrations (10 ng/ml) of phorbol myristate acetate (PMA) and Pgp expression was analyzed by FACScan. Such an activation resulted in an augmentation of Pgp in T cells (Figure 2).

Table 2. Effect of actinomycin D and cycloheximide on P-glycoprotein expression.

Experimental Conditions	Pgp Positive CD3+ Cells (%)
Experiment # 1	
MNC + PHA	38.0
MNC + PHA + Actinomycin D (1 μg/ml)	5.0
Experiment # 2	
MNC + PHA	82.1
MNC + PHA + Cycloheximide (10 μM)	15.5
MNC + PHA + Cycloheximide (20 μM)	9.3
MNC + PHA + Cycloheximide (30 μM)	3.8

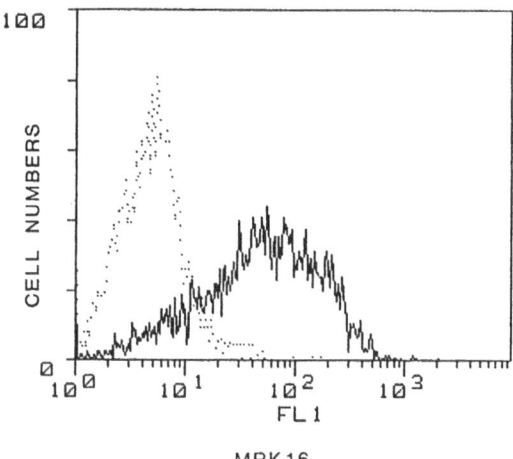

MRK 16

Figure 2. Effect of PKC activation with low concentrations of PMA on Pgp expression. Dotted line shows Pgp expression in the absence and the solid line in the presence of PMA.

Secondly, depletion of PKC by prolonged (18 hours) exposure of MNC to high concentration of PMA (162nM) (a process known to degrade PKC),[13] followed by activation with PHA for 48 hours, resulted in an inhibition of PHA-induced Pgp augmentation (Figure 3). The inhibition of Pgp expression was associated with downregulation of both PKCα and PKCβ isoforms (Table 3 and Figure 4).

Thirdly, PKC inhibitors staurosporin and H7, in a dose-dependent manner, inhibited PHA-induced increase in Pgp expression (Table 4). These data demonstrate a role of PKC in an augmentation of Pgp expression following activation of MNC with PHA.

Table 3. Effect of prolonged exposure of PMA on PKC isoforms in PHA-activated MNC.

Exp. Conditions	PKC Isoforms + Cells (%)		Density of PKC Isoforms (MFC#)	
	PKCα	PKCβ	PKCα	PKCβ
MNC + PHA	16.2	65.2	7.84	12.53
MNC + PMA + PHA	3.7	35.1	6.40	9.50

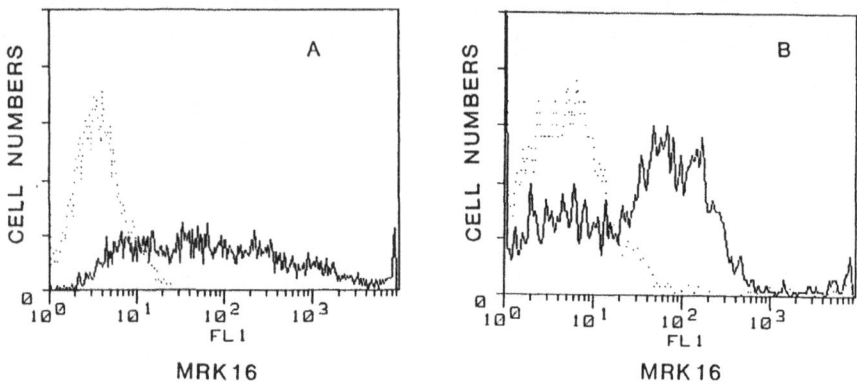

Figure 3. Effect of PKC depletion on PHA-induced augmentation of Pgp. Solid lines show Pgp expression following PHA stimulation with (B) or without (A) prolonged exposure to PMA.

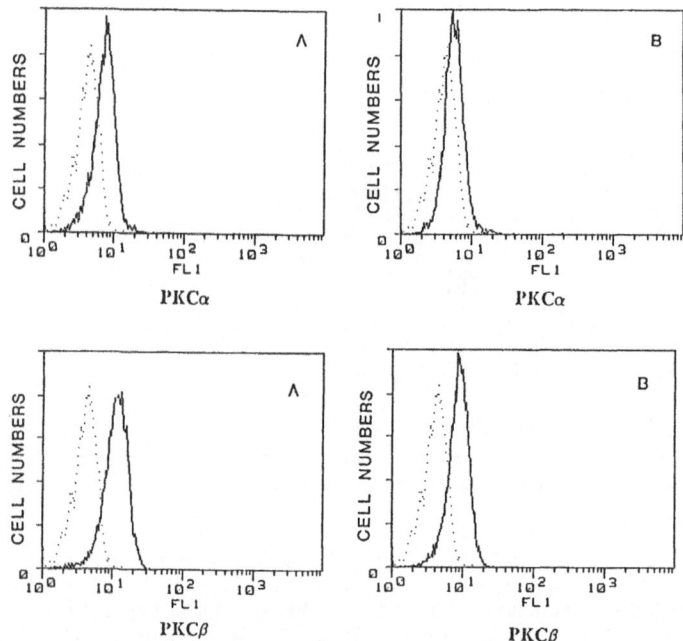

Figure 4. Effect of prolonged and high dose treatment of PMA on PKC isoforms

Table 4. Effect of staurosporin and H7 on PHA-induced Pgp expression.

Exp. Conditions	P-Glycoprotein + Cells (%)
MNC + PHA	67.3 ± 7.1
MNC + PHA + Staurosporin (2.5 nM)	67.1 ± 2.4
MNC + PHA + Staurosporin (5.0 nM)	24.0 ± 2.6
MNC + PHA + Staurosporin (10.0 nM)	13.3 ± 2.0
MNC + PHA + H7 (25 µM)	$37.0 + 2.0$

Cyclosporin A Inhibits Pgp Expression in T Cells

Mononuclear cells were stimulated with PHA for 48 hours in the presence or absence of various concentrations of CsA. Cells were washed and analyzed for Pgp expression with MRK16 monoclonal antibody, using FACScan. Cyclosporin A inhibited Pgp expression (Table 5 and Figure 5B). No cell death was observed at any of the concentrations of CsA tested.

Table 5. Effect of cyclosporin A on Pgp expression.

Cyclosporin A (μg/ml)	Percent Pgp+ Cells
None	70.0
0.125	68.0
0.25	55.6
0.50	0.0
1.00	4.0

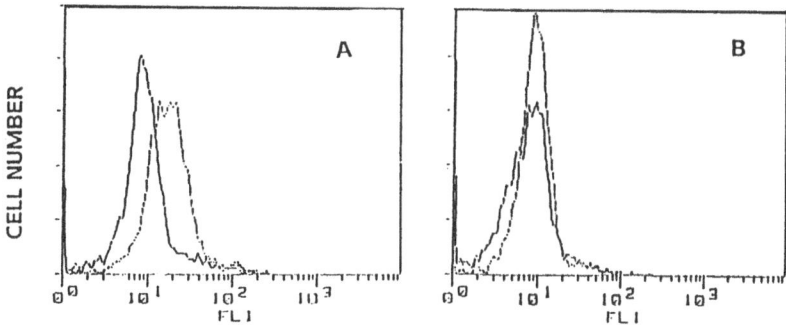

Figure 5. Effect of CsA (1 μg/ml) on PHA-induced P-glycoprotein expression

The inhibition of Pgp expression by CsA was not due to decrease in cell size (mean channel number [MC#]) but was associated with an inhibition of hyperpolarization of plasma membrane as demonstrated by decreased DiOC5 dye uptake in the presence of CsA (Table 6).

Table 6. Effect of CsA on plasma membrane potential and cell size

Exp. Conditions	Plasma Membrane Potential (MC#)	Cell Size
MNC	26 ± 0	64 ± 1
MNC + PHA	100 ± 4	129 ± 1
MNC + PHA + CsA (1 μg/ml)	77 ± 5	141 ± 2

Cyclosporin A-Mediated Inhibition of Pgp is Associated with Downregulation of PKC

Recently, Walker et al[14] have shown that CsA inhibits PKC in renal capillary endothelium. Therefore, we investigated whether the inhibition of Pgp expression by CsA was associated with downregulation of PKC isforms. Mononuclear cells were stimulated with PHA in the presence or absence of 1 μg/ml of CsA for 48 hours and the expression of PKC isoforms was measured with anti-PKCα and anti-PKCβ monoclonal antibodies and FACScan. Data are expressed both as the percent positive cells and the amount of PKC (as demonstrated by mean fluorescence channel numbers [MFC#]). Data in Figure 6 and Table 7 show that CsA inhibited both the number of cells expressing PKC isoforms and the density of PKC isoforms. Although no difference was observed between proportions of cells showing downregulation of either isoform (50%), the density of PKCα isoform was decreased to a greater extent (60%) as compared to that of PKCβ isoform (30%).

Table 7. Effect of cyclosporin A on PKC isoforms.

| Exp. Conditions | Percent Positive Cells | | Mean Fluorescence Channel # | |
	PKCα	PKCβ	PKCα	PKCβ
MNC + PHA	50	44	95	66
MNC + PHA + CsA	25	22	44	44

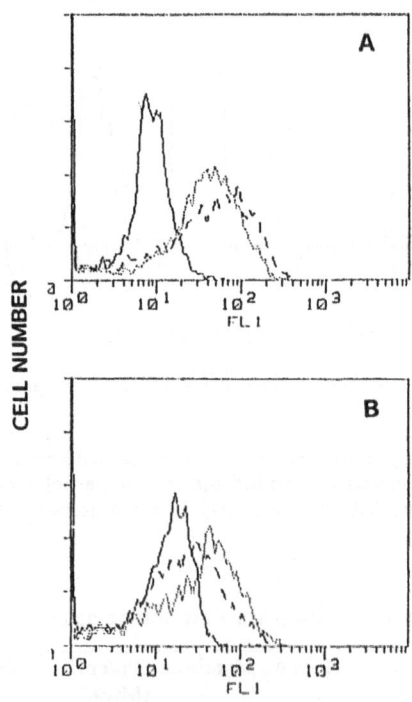

FLUORESCENCE INTENSITY

Figure 6. Effect of CsA on PKC isoforms. CsA inhibited both PKCα (A) and PKCβ (B) isoforms. Closed dotted lines show PKC isoform in the presence of CsA and the wide interrupted lines in the absence of CsA. Solid line shows background control.

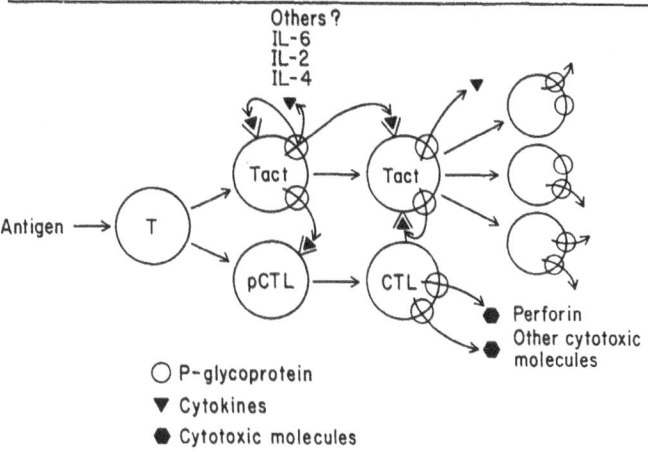

Figure 7. Proposed model of P-glycoprotein function in human T cell functions.

DISCUSSION

P-glycoprotein is a member of a highly conserved multigene family with three genes in rodents and two genes in primates.[15] Transfection studies have shown that only class I Pgp (coded by *mdr 1* gene) in humans confers the multidrug resistant phenotype in cancer cells.[16-18] The physiologic role of Pgp in presently unclear. Previous studies with monoclonal antibodies against Pgp have failed to demonstrate the presence of Pgp in thymus, lymph nodes and spleen cells.[19-23] Recently several investigators have demonstrated the presence of low levels of *mdr 1* mRNA in peripheral blood MNC and the bone marrow cells from patients with leukemia.[24,25] More recently, both *mdr 1* mRNA and membrane Pgp expression have been reported in human bone marrow stem cells.[26] We have shown that low levels of *mdr 1* mRNA and membrane Pgp are expressed in freshly isolated peripheral blood T cells.[11] The activation of MNC with PHA resulted in an augmentation of both *mdr 1* mRNA and membrane Pgp and the expression of Pgp was preferential in CD8+ T cells.[11] The PHA-induced augmentation of Pgp expression appears to occur primarily in G1-S phase of cell cycle. It is interesting to note that the cystic fibrosis gene encoded protein, the cystic fibrosis transmembrane conductance regulator (CFTR), that shows strong homology with Pgp[27] and is a Cl⁻ channel,[28] also shows Cl⁻ permeability largely restricted to G1-S phase of cell cycle.[29] Augmentation of Pgp requires both transcriptional and translational events.

Recent molecular cloning and biochemical characterization of PKC have indicated the presence of at least 7 isoforms of PKC in mammalian tissue, of which the best characterized are α, β, and γ isoforms.[30] In T cell lines only PKCα and PKCβ have been observed.[31] In this study we have demonstrated that that both PKCα and β isoforms, albeit PKCα to a greater extent, are involved in PHA-induced augmentation of Pgp.

We have also shown that Pgp in T cells is a metabolically active efflux pump.[11] Furthermore, we have reported that anti-Pgp monoclonal antibody (MRK16), in a concentration-dependent manner, inhibits effector CTL function.[11] We propose that Pgp in CD8+ T cells plays a physiological role in cytotoxic T cell effector function by transporting cytotoxic molecules from effector cells to target cells. It is also likely that Pgp protects CD8+ T cells from the toxic effects of its own cytotoxic molecule formed during an immune response by transporting them out of the cells. The role of Pgp in subset of CD4+ T cells remains to determined. We hypothesize that certain cytokines, with negative charge and hydrophobicity, might be the substrate for Pgp and therefore, Pgp will participate in their secretion from the cells. Furthermore, it is also likely that Pgp in CD4+ T cells is restricted to CD4+ cytotoxic T cells. A hypothetical model for the role of Pgp in T cell physiology is shown in Figure 7. Cyclosporin A is known to inhibit T cell activation.[32] In this study we show that CsA in a dose-dependent manner inhibits PHA-induced amplification of Pgp in MNC. The inhibition of Pgp expression was associated with an inhibition of PHA-induced hyperpolarization of plasma membrane but not associated with any decrease in cell size. Furthermore, inhibition of Pgp by CsA was associated with downregulation of both PKCα and PKCβ isoforms. The CsA induced inhibition of Pgp expression appears to be at post-transcription/ post-translational level, as no effect of CsA was observed on *mdr 1* mRNA (Gupta, Kim and Gollapudi-- unpublished data).

ACKNOWLEDGEMENT

The part of the work reported here was supported by USPHS grant AI-26456

REFERENCES

1. P. Gros, J. Croop, and D. Housman, Mammalian multidrug resistance gene: complete cDNA sequence indicates strong homology to bacterial transport proteins, *Cell* 47:371 (1986).
2. C.-J. Chen, J.E. Chin, K. Ueda, D.P. Clark, I. Pastan, and M.M. Gottesman, Internal duplication and homology with bacterial transport proteins in the *mdr 1* (P-glycoprotein) gene from multidrug resistant human cells, *Cell* 47:381 (1986).
3. J.H. Gerlach, J.A. Endicott, P.F. Juranka, G. Henderson, F. Sarangi, K.L. Deuchars, and V. Ling, Homology between P-glycoprotein and a bacterial hemolysin transport protein suggests a model for multidrug resistance, *Nature* 324:485 (1986).
4. M.M. Gottesman M.M. and I. Pastan, The multidrug transporter, a double-edge sword, *J. Biol. Chem.* 263:12163 (1988).
5. J.M. Croop, P. Gros, and D.E. Housman, Genetics of multidrug resistance, *J. Clin. Invest.* 81:1303, 1988.
6. P.F. Juranka, R.L. Zastawny, and V. Ling, P-glycoprotein: multidrug resistance and a superfamily of membrane associated transport protein, *FASEB J.* 3:2583 (1989).
7. S.J. Foote, J.K. Thompson, A.F. Cowman, and D.J. Kemp, Amplification of the multidrug resistant gene in some chloroquine-resistant isolates of *P. falciparum. Cell* 57:921 (1989).
8. C.M. Wilson, A.E. Serrano, A. Wesley, M.P. Bogenschultz, A.H. Shanker, and D.F. Wirth, Amplification of a gene related to mammalian *mdr* genes in drug-resistant *Plasmodium falciparum, Science* 244:1184 (1989).
9. S. Gollapudi S. and S. Gupta, Human immunodeficiency virus-1 (HIV-1)-induced P-glycoprotein expression, *Biochem. Biophysic. Res. Commun.* 171:1002, 1990.
10. H. Hamada and T. Tsuruo, Functional role for the 170-180 kda glycoprotein specific to drug resistant tumor cells as treated by monoclonal antibodies, *Proc. Natl. Acad. Sci. (USA)* 83:7785 (1986).
11. S. Gupta, C.H. Kim, T. Tsuruo, and S. Gollapudi, Preferential expression of multidrug resistant gene 1 product (P-glycoprotein), a functionally active efflux pump, in human CD8+ T cells. A role in cytotoxic effector function, (submitted).
12. Y. Nishizuka, The role of protein kinase C in cell surface signal transduction and tumor promotion, *Nature* 308:693 (1984).
13. S. Young, P. Parker, A. Ulrich, and S. Stable, Down regulation of protein kinase C is due to an increased rate of degradation, *Biochem. J.* 244: 775 (1987).
14. R.J. Walker, V.A. Lazzaro, G.G. Duggin, J.S. Horvath, and D.J. Tiller. Cyclosporin A inhibits protein kinase C activity; a contributing mechanism in the development of nephrotoxicity. *Biochem. Biophys. Res. Commun.* 160: 409 (1989).
15. W.F. Ng, F. Sarangi, R.L. Zastawny, V. Veinot-Drebot, and V. Ling, Identification of members of the P-glycoprotein multigene family, *Mol. Cell. Biol.* 9:1224 (1989).
16. K. Ueda, C. Cardarelli, M.M. Gottesman, and I. Pastan, Expression of a full length cDNA

human *MDR 1* gene confers resistance to colchicine, doxorubicin, and vinblastin, *Proc. Natl. Acad. Sci. (USA)* 84:3004, (1987).

17. P.G. Dabenham, N. Kartner, L. Siminovitch, J.R. Riordan, and V. Ling, DNA-mediated transfer of multiple drug resistance and plasma membrane glycoprotein expression, *Mol. Cell. Biol.* 2:881 (1982).

18. D. Shen, A. Fojo, I. Roninson, J. Chin, R. Soffir, I. Pastan, and M.M. Gottesman, Multidrug resistance of DNA-mediated transformants is linked to transfer of the human mdr gene, *Mol. Cell. Biol.* 6:4039, (1986).

19. T. Thiebaut, T. Tsuruo, H. Hamada, M.M. Gottesman, I. Pastan, and M.C. Willingham, Cellular localization of multidrug resistant gene product P-glycoprotein in normal human tissues, *Proc. Natl. Acad. Sci. (USA)* 84:7734, (1987).

20. J.M. Croop, M. Raymond, D. Haber, A. Devault, R.J. Arceci, P. Gros, and D.E. Houseman, The mouse multidrug resistance (mdr) genes are expressed in a tissue specific manner in normal mouse tissues, *Mol. Cell. Biol.* 9:1346 (1989).

21. E. Georges, G. Bradley, J. Gariep, and V. Ling, Detection of P-glycoprotein isoforms by gene-specific monoclonal antibodies, *Proc. Natl. Acad. Sci. (USA)* 87:152 (1990).

22. C. Cordon-Cardo, J.P. O'Brien, D. Boccia, D. Casals, J.R. Bertino, and M.R. Melamed, Expression of the multidrug resistance gene product (P-glycoprotein) in human normal and tumor cells, *J. Histochem. Cytochem.* 28:1277 (1990).

23. I. Sugawara, I. Kataoka, Y. Morishita, H. Hamada, T. Tsuruo, S. Itoyama, and S. Mori, Tissue distribution of P-glycoprotein encoded by a multidrug resistant gene as revealed by a monoclonal antibody, MRK16, *Cancer Res.* 48:1926, (1988).

24. K.E. Noonen, C. Beck, T.A. Holzmayer, J.E. Chin, J.S. Wunder, I.L. Andrulis, A.F. Gazdar, C. L. Willman, B. Griffith, D.D. Von Hoff, and I.B. Roninson, Quantitative analysis of MDR 1 (multidrug resistance) gene expression in human tumors by polymerase chain reaction, *Proc. Natl. Acad. Sci. (USA)* 87:7160 (1990).

25. R. Pirker, L.J. Goldstein, H. Ludwig, W. Linkesh, C. Lechner, M.M. Gottesman, and I. Pastan, Expression of a multidrug resistant gene in blast crisis of chronic myelogenous leukemia, *Cancer Commun.* 1:141 (1989).

26. P.M. Chaudhary and I.B. Roninson, Expression and activity of P-glycoprotein, a multidrug efflux pump, in human hematopoietic stem cells, *Cell* 66:85 (1991).

27. J.R. Riordan, J.M. Rommens, B.-S. Kerem, N. Alon, R. Rozmahel, Z. Grzelczak, J. Zielenski, S. Lok, N. Plavsic, J. Chou, M.L. Drumm, M.C. Iannuzzi, F.S. Collins, L. Tsui, Identification of cystic fibrosis gene: cloning and characterization of complementary DNA. *Science* 245:1066 (1989).

28. M.P. Anderson, R.J. Gregory, S. Thompson, D.W. Souza, S. Paul, R.C. Mulligan, A.E. Smith, and M.J. Welsh, Demonstration of CFTR is a chloride channel by alteration of its ionic selectivity, *Science* 253:202 (1991).

29. J.K. Rubein, K.L. Kirk, T.A. Rado, and R.A. Frizzell, Cell cycle dependence of chloride permeability in normal and cystic fibrosis lymphocytes, *Science* 248:1416 (1990).

30. Y. Nishizuka, The molecular heterogeneity of protein kinase C and its implications for cellular regulation, *Nature* 334:661 (1988).

31. C.A. Koretzky, M. Wahi, M.E. Newton, and A. Weiss, Heterogeneity of protein kinase C isoenzyme gene expression in human T cell lines. Protein kinase Cβ is not required for certain functions, *J. Immunol.* 143:1692 (1989).

32. E.M. Shevach, The effect of cyclosporin A on the immune system, *Ann. Rev. Immunol.* 3:397 (1985).

INTEGRINS, γδ T CELLS, AND AUTOIMMUNITY

Ethan M. Shevach

Laboratory of Immunology
National Institute of Allergy and Infectious Diseases
National Institutes of Health
Bethesda, MD 20892

INTRODUCTION

T lymphocytes can be divided into two major lineages. The majority of peripheral T lymphocytes express the αβ TCR and recognize antigen in association with MHC class I or class II molecules, while a second minor subpopulation of T lymphocytes expresses the γδ chains (1). The nature of the antigen recognized by this latter population or the nature of the restriction element utilized remain poorly characterized (2). One major subpopulation of γδ cells which express the Vγ1.1Cγ4, Vδ6Cδ TCR (hereafter referred to as Vγ1.1Vδ6 TCR) has been isolated from newborn thymus and hybridomas which express this TCR spontaneously produce IL-2 (3). It has been postulated that spontaneous cytokine production by these hybridomas is secondary to recognition of an autoantigen, possibly an autologous heat shock protein, expressed on the surface of the thymic hybridomas themselves (4).

We have studied in detail the cellular and molecular properties of a similar population of T cells which express the Vγ1.1Vδ6 TCR which were originally isolated from murine dendritic epidermal T cells (5-8). Although these cells also secrete cytokines spontaneously, activation of these cells appears to be a very complex process which involves not only recognition of an autoantigen, but also interaction of the vitronectin receptor (VNR) expressed on the surface of these cells with its ligand in extracellular matrix (ECM-) proteins (9). In the present report, I will review in depth the functional properties of this unique subpopulation of T cells which express the γδ TCR and will discuss their possible role in both normal T cell development and in the pathogenesis of autoimmune diseases.

THE ROLE OF THE VNR AS AN ACCESSORY MOLECULE FOR ACTIVATION OF Vγ1.1Vδ6 T CELLS

During the course of studies designed to evaluate cytokine production by γδ T cell lines derived from murine epidermis, we noted (5) that all cell lines which expressed the Vγ1.1Vδ6 TCR, but not the more common Vγ3Vδ1 TCR, produced cytokines spontaneously in culture in the absence of stimulation by an exogenous ligand (Table 1). When anti-TCR reagents were added to these lines, marked inhibition of cytokine production was observed. These results strongly suggested that the anti-TCR reagents were blocking antigen recognition and that spontaneous cytokine production was secondary to recognition of an antigen expressed in the culture medium or on the surface

Mechanisms of Lymphocyte Activation and Immune Regulation IV: Cellular Communications
Edited by S. Gupta and T.A. Waldmann, Plenum Press, New York, 1992

49

Table 1. Constitutive cytokine production by a Vγ1.1Vδ6 T cell hybridoma

Medium	Anti-CD3	Anti-clonotype
40,016	753	13,149

Results are expressed and cpm ^{3}H-TdR incorporation by an indicator line.

of the hybridoma cells themselves. When these cells were adapted to serum-free medium, spontaneous cytokine production was completely inhibited. Although this result again suggested that the "antigen" recognized by the Vγ1.1Vδ6 TCR was a component of the fetal calf serum used in the tissue culture medium, activation of cytokine production in serum-free medium could be induced by the addition of the RGD-containing ECM-proteins fibronectin, vitronectin, and fibrinogen (Table 2). Furthermore, the response to the ECM-proteins could be completely inhibited by a monoclonal antibody (mAb) to the murine VNR (9). However, cytokine production could also be inhibited by anti-TCR mAbs.

Table 2. Induction of cytokine production by ECM-proteins

Culture Conditions	Inhibitor	IL-4 Production (U/ml)
Serum-free	None	0
Fibronectin	None	588
Fibronectin	RGDS	<1
Fibronectin	RGES	500
Fibronectin	Anti-VNR	0
Fibronectin	Anti-CD3	0

One possible explanation for the dual inhibitory effects of anti-TCR and anti-VNR reagents was that the Vγ1.1Vδ6 TCR recognized an non-RGD site on the ECM-protein as its ligand and that engagement of this site by the TCR together with engagement of the RGD site by the VNR were both required for spontaneous cytokine production. However, it is very unlikely that sites on the ECM-proteins were serving as the ligand for the γδ TCR as fibronectin, fibrinogen, and vitronectin share minimal sequence homology other than RGDS. Furthermore, the ability of a synthetic RGDS-containing peptide (PepTite 2000) to activate cytokine production is most consistent with the view that RGDS is both necessary and sufficient for both cell adhesion and cell activation. We regard it as very unlikely that both the VNR and the TCR could both recognize the RGDS sequence.

THE ROLE OF THE AUTOREACTIVE Vγ1.1Vδ6 TCR IN T CELL ACTIVATION

Although recognition of the RGDS sequence by the VNR was sufficient to induce cytokine production, the TCR complex also played a role in the activation of the Vγ1.1Vδ6 T cells as mutants of one of the autoreactive cell lines that failed to express the Vγ1.1Vδ6 TCR secondary to a selective loss of TCR γ-chain expression failed to respond to stimulation by serum or RGDS-containing proteins. One possible explanation for this finding is that activation of T cells via the VNR is similar to activation mediated by other cell surface antigens which are capable of mediating alternative pathways of T cell activation. Thus, stimulation of murine T cells via the Thy-1 or Ly-6 (10) molecules or human T cells via the CD2 molecule (11) requires coexpression of the TCR, but engagement of the TCR by its ligand is not required. However, the VNR is expressed on almost all T cell hybridomas which express either the αβ or the γδ TCR, yet spontaneous cytokine production has only been observed on cell lines or hybridomas which express the Vγ1.1Vδ6 TCR.

An alternative explanation for these experimental observations is that co-expression of the VNR and the TCR is required for activation of cytokine production, and that both the VNR and the TCR must be engaged by their respective ligands. The most likely, if not only candidate, for a ligand for the TCR is an antigen expressed on the surface of the hybridoma cells themselves. We have tested this hypothesis by taking advantage of the observations of O'Brien et al (3) that Vγ1.1 T cell hybridomas derived by fusing newborn thymocytes to BW5147 spontaneously secrete IL-2. Two different selection techniques were used (6). In one of the screening studies, hybridomas were selected solely based on their capacity to spontaneously produce cytokines when their supernatants were tested on cell lines capable of detecting IL-2, IL-3, IL-4, or GM-CSF. After spontaneous producers were identified, biochemical and molecular studies of their TCR were performed (7). All of the spontaneous producers were shown to express the Vγ1.1 TCR. Furthermore, all of the hybridomas derived from these fusions phenotypically resembled the spontaneous cytokine producers derived from murine epidermis as cytokine production was inhibited by anti-TCR and anti-VNR reagents (Table 3). Similar conclusions were drawn from the second type of screening study in which hybridomas were selected solely based on the expression of the αβ or the γδ TCR by FACS analysis and then subsequently examined for spontaneous cytokine production and the molecular properties of the TCR. The only spontaneous cytokine producer derived from this screen also expressed the Vγ1.1 TCR and resembled all the other hybridomas isolated from newborn thymus (Table 3, hybridoma GD73). As the frequency of spontaneous cytokine producing hybridomas (1/35) was similar using these two screening assays, these data are strong evidence in favor of a selective association between spontaneous cytokine production and expression of the Vγ1.1 TCR.

Table 3. Cytokine production by Vγ1.1Vδ6 thymocyte hybridomas

Hybridoma	γ-chain	δ–chain	Media	Anti-CD3	Anti-VNR
AA37	Vγ1.1	Vδ6	89,746	6,462	19,780
BB27	Vγ1.1	Vδ6	194,083	22,348	11,545
CC48	Vγ1.1	Vδ6	146,100	11,611	10,395
EE32	Vγ1.1	Vδ6	198,843	11,963	12,310
FF3	Vγ1.1	Vδ4	163,956	27,149	10,820
GD73	Vγ1.1	Vδ6	156,243	26,683	30,092

Results are expressed as cpm ^{3}H-TdR incorporation by an indicator cell line.

TRANSFECTION STUDIES WITH THE Vγ1.1Vδ6 TCR

Collectively, the studies described above are consistent with the hypothesis that constitutive cytokine production by Vγ1.1Vδ6 T cells is secondary to recognition of a self-antigen by the γδ TCR. In order to directly address the issue of the role of the Vγ1.1Vδ6 TCR, the VNR, and the self-antigen in the process of T cell activation, we have transfected cDNAs encoding the Vγ1.1Cγ and the Vδ6Cδ chains of the T cell hybridoma, T195, into both human and murine cell lines (8). In our initial studies, we used as a recipient cell line, TG40, a derivative of the murine cytochrome c-specific T cell hybridoma, 2B4, which fails to express either the αβ or the γδ TCR as determined both by FACS analysis and mRNA studies. We isolated one clone, TG524, which expressed the Vγ1.1Vδ6 TCR as determined by FACS with an anti-clonotypic mAb and which was also reactive with an anti-pan γδ mAb and with anti-CD3. TG524 and TG40 were also reactive with the anti-VNR reagents as determined by flow cytometry. When TG524 was evaluated for cytokine production, it spontaneously produced IL-2 which is the cytokine produced by the 2B4 hybridoma, while no IL-2 production could be detected when either TG40 or 2B4 were cultured under identical condition (Table 4, Exp.I). Thus, spontaneous IL-2 production is secondary to expression of the Vγ1.1Vδ6 TCR which must recognize the autoantigen expressed by TG40 cells themselves. Furthermore, TG524 appeared to phenotypically resemble T195 in all its activation properties. Thus, spontaneous IL-2 production could be inhibited by anti-TCR and anti-VNR reagents in a manner identical to that seen with the T195 or the thymocyte hybridoma lines (Table 5). In contrast, the anti-VNR reagents failed to inhibit IL-2 production in response to antigen and antigen presenting cells by the

Table 4. Cytokine production by TG524 and J1120.

	Stimulus	TG524	TG40	2B4
Exp. I	Medium	37,263	471	521
	Anti-CD3	64,606	448	53,975
	Stimulus	J1120	T3.1	JE6.1
Exp.II	PMA	4,392	2,445	5,300
	PMA + Anti-CD3	39,054	4,841	67,805
	PMA + Anti-clono	65,271	6,101	4,943

Results are expressed as cpm ^{3}H-TdR incorporation by an indicator cell line.

Table 5. Inhibition of constitutive IL-2 production by TG524.

Medium	Anti-CD3	Anti-clono	Anti-VNR	RGDS	RGES
12,070	1,269	3,410	1,822	1,416	12,931

Results are expressed as cpm ^{3}H-TdR incorporation by an indicator cell line.

parental 2B4 cell line even though 2B4 expressed the VNR at a level identical to that of TG524.

We also transfected the TCR α-chain loss mutant,T3.1, of the human Jurkat leukemia cell line. One clone, J1120, was isolated which expressed the murine Vγ1.1Vδ6 TCR in association with the human CD3 complex. However, in contrast to TG524, J1120 did not demonstrate spontaneous cytokine production even in the presence of PMA as a costimulator (Table 4, Exp.II). The murine Vγ1.1Vδ6 TCR was capable of signal transduction as IL-2 could readily be detected following stimulation of J1120 with anti-murine pan γδ, anti-clonotype, or anti-human CD3. The simplest explanation for this observation is that Jurkat cells, in contrast to TG524, do not express the autoantigen. In preliminary studies, we have shown that Jurkat cells express the VNR although we have not yet evaluated the function of this integrin in detail.

One of the differences noted between TG524 and T195 was that TG524 did not spontaneously secrete cytokines at maximal levels as higher levels of IL-2 were routinely detected when TG524 was stimulated with plate-bound anti-CD3 or PMA/ionophore. We therefore had the opportunity to directly evaluate whether we could enhance IL-2 production by co-culture of TG524 with cell lines that were likely to express the autoantigen (Table 6). As the Jurkat transfectant, J1120, did not secrete cytokines spontaneously, we could co-culture J1120 with a variety of murine cell types (including T195 itself) which might express the autoantigen. The results of these co-culture experiments with either the murine or human transfectant were identical. Under no conditions could IL-2 production by TG524 be augmented during the co-culture. Similarly, we never observed IL-2 production by J1120 when it was cultured with mouse cells. The results with J1120 can readily be explained by the absence of expression of an accessory molecule required for the interaction of the human cell with the murine stimulator. However, other groups have demonstrated that Jurkat cells transfected with either the murine αβ (12) or the γδ (13) TCR can respond to murine stimulator cells. Of course, different subpopulations of cells and the receptors expressed by them may have different requirements for the use of accessory molecules. Our failure to see augmentation when TG524 was co-cultured with different cell populations which express the purported autoantigen is problematic and may be secondary to the unique requirements of the VNR as an accessory molecule on this cell.

A THE TCR AND THE SELF-ANTIGEN INTERACT ON THE SURFACE
OF THE SAME CELL

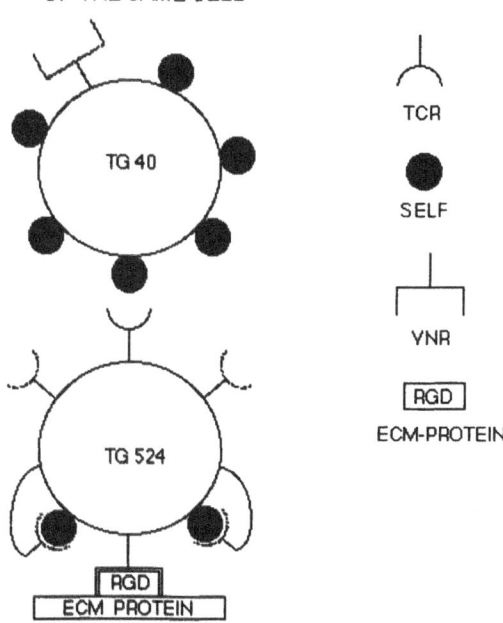

B TCR + VNR MUST BE ENGAGED BY LIGANDS ON THE SAME SURFACE

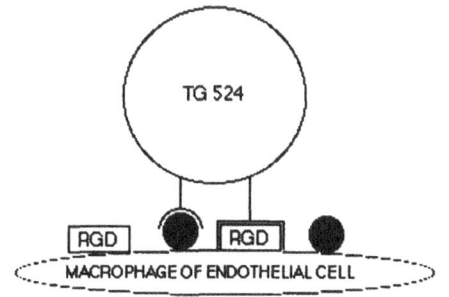

Figure 1. What is the mechanism of integrin facilitated auto-antigen presentation?

Table 6. Co-culture studies with Vγ1.1Vδ6 transfectants.

	Responder	Stimulus	^{3}H-TdR Incorporation
Exp. I	TG524	Medium	13,156
	TG524	Anti-CD3	59,574
	TG524	TG40	11,919
Exp.II	J1120	PMA	4,392
	J1120	PMA+Anti-γδ	51,641
	J1120	PMA+T195	2,235

CONCLUSIONS AND SPECULATIONS

Collectively, these data offer strong evidence in support of the hypothesis that the Vγ1.1Vδ6 TCR reacts with an autoantigen. The nature of this autoantigen is completely unknown although studies by other groups (4) have raised the possibility that the antigen may be an autologous heat shock protein. One of the major questions that must be addressed in the future is what is the role of this autoreactive population of T cells in vivo. First, one must question whether an autoreactive cell can be allowed to exist in an unregulated manner in vivo. It is quite possible that the presence of such a population of cells in the thymus and in the peripheral lymphoid tissues might lead to the development of autoimmune disease. Although evidence that self-reactive γδ T cells are deleted (14) or exhibit clonal anergy (15) has been derived from studies with transgenic mice, there is little solid data that γδ T cells although they appear to pass through the thymus are either positively or negatively selected in the thymus. One unique way in which the activity of the Vγ1.1Vδ6 population might be controlled might be to regulate the expression of the VNR which functions as a requisite accessory molecule for activation of these cells. Indeed, when the autoreactive hybridomas are cultured in serum-free medium, cytokine production is abrogated even though the γδ TCR should be fully capable of engaging the autoantigen. Thus, the affinity of this TCR for the autoantigen is low and must be enhanced by interaction of the VNR with its ligand in ECM-proteins. The VNR is an unusual cell surface antigen as our studies have demonstrated (9) that its expression is only induced following prolonged (>7 days) T cell activation in vitro and we have not yet defined conditions for the induction of its expression in vivo. During the course of T cell differentiation in vivo in the thymus or during the course of an acute inflammatory response, the Vγ1.1Vδ6 TCR may engage the self-antigen with a low affinity; following a period of prolonged receptor occupancy, expression of the VNR may result with consequent high affinity interaction of the TCR with the autoantigen. It is only at this point in time that the autoreactive T cells will produce cytokines that may be beneficial or deleterious to the function of surrounding tissues and cells.

One additional regulatory aspect of the function of the Vγ1.1Vδ6 T cells may be derived from our co-culture studies. It appears that this population of cells is incapable of recognizing the autoantigen expressed in a "trans" configuration on the surface of another cell type. We have therefore postulated that these cells present the autoantigen to their own TCR on the same cell surface (Figure 1A). Indeed, the requirement for engagement of the VNR by ECM-proteins on the surface of the plastic tissue culture dish may dictate this form of autopresentation as the VNR may be cleared from the face of the cell which is nonadherent to the plastic. Thus, signal transduction requires not only both the engagement of the TCR and VNR by their respective ligands, but also their physical apposition. If a similar apposition of TCR and VNR occurs in vivo, this would mandate that the TCR recognize an autoantigen on the surface of the same cell which expresses an ECM-protein that might be engaged by the VNR (Figure 1B). This model of T cell activation would further restrict activation of the autoreactive cells to unique APC or to sites of inflammation where ECM-proteins would be expressed on the surface of macrophages or endothelial cells.

REFERENCES

1. D.H. Raulet, The structure, function, and molecular genetics of the γδ T cell receptor, *Annu. Rev. Immunol.*, 7: 175 (1989).
2. J.A. Bluestone, R.Q. Cron, T.A. Barrett, B. Houlden, A.I. Sperling, A. Dent, S. Hedrick, B. Rellahan, and L.A. Matis, Repertoire development and ligand specificity of Murine TCRγδ cells, *Immunol. Rev.*, 120:5 (1991).
3. R.L. O'Brien, M.P. Happ, A. Dallas, E. Palmer, R. Kubo, and W.K. Born, Stimulation of a major subset of lymphocytes expressing T cell receptor γδ by an antigen derived from Mycobacterium tuberculosis, *Cell*, 57: 667 (1989).
4. W. Born, L. Hall, A. Dallas, J. Boymel., T. Shinnick, D. Young, P. Brennan, and R.O. O'Brien, Recognition of a peptide antigen by heat shock reactive γδ T lymphocytes, *Science*, 249:67 (1990).
5. K. Roberts, W.M. Yokoyama, P.J. Kehn, and E.M. Shevach, The vitronectin receptor serves as an accessory molecule for the activation of a subset of γδ T cells, *J. Exp. Med.*, 173: 231 (1991).
6. D.B. Wilde, K. Roberts, K. Sturmhofel, G. Kikuchi, J.E. Coligan, and E.M. Shevach. Mouse autoreactive γδ T cells. I. Functional properties of autoreactive T cell hybridomas, *Eur. J. Immunol.*, In Press.
7. A. Ezquerra, D.B.Wilde, T.J. McConnell, K. Sturmhofel, R.B. Valas, E.M. Shevach, and J.E. Coligan, Autoreactive γδ T cells II. Molecular characterization of the T cell receptor, *Eur. J. Immunol.*, In Press.
8. G.E. Kikuchi, K. Roberts, E.M. Shevach, and J.E. Coligan, Gene transfer demonstrates that the Vγ1.1Cγ4V86C8 TCR is essential for autoreactivity, *J. Immunol.*, In Press.
9. S.R. Maxfield, K. Moulder, F. Koning, A. Elbe, G. Stingl, J.E. Coligan, E.M. Shevach, and W.M. Yokoyama, Murine T cells express a cell surface receptor for multiple extracellular matrix proteins: identification and characterization with monoclonal antibodies, *J. Exp. Med.* 169: 2173 (1989).
10. J.J. Sussman, T. Saito, E.M. Shevach, R.N. Germain, and J.D. Ashwell, Thy-1 and Ly-6 mediated lymphokine production and growth inhibition of a T cell hybridoma require coexpression of the T cell antigen receptor complex, *J. Immunol.*, 140:2520 (1988).
11. J.B. Breitmeyer, J.F. Daley, H.B. Levine, and S.F. Schlossman, The T11 (CD2) molecule is functionally linked to the T3/Ti T cell receptor in the majority of T cells, *J. Immunol.*, 139: 2899 (1987).
12. T. Saito, A. Weiss, J. Miller, M.A. Norcross, and R.N. Germain, Specific antigen-Ia activation of transfected T cells expressing murine Ti αβ-human T3 receptor complexes, *Nature*, 325: (1987).
13. W.L. Havran, Y. H. Chien, and J. P. Allison, Recognition of self antigens by skin derived T cells with invariant γδ antigen receptors, *Science*, 252:1430 (1991).
14. A.L.Dent, L.A. Matis, F. Hoshmand, S.M. Widacki, J.A. Bluestone, and S.M. Hedrick, Self-reactive γδ T cells are eliminated in the thymus, *Nature*, 343:714 (1990).
15. M. Bonneville, I. Ishida, S. Itohara, S. Verbeek, A. Berns, O. Kanagawa, W. Haas, and S. Tonegawa, Self-tolerance to transgenic γδ T cells by intrathymic inactivation, *Nature*, 344:163 (1990).

THE INTERLEUKIN-2 RECEPTOR: A TARGET FOR IMMUNOTHERAPY

Thomas A. Waldmann, Carolyn Goldman, Lois Top, Angus
Grant, Jack Burton, Richard Bamford, Erich Roessler, Ivan
Horak, Sara Zaknoen, Claude Kasten-Sportes, Jeffrey White,
and David Nelson

Metabolism Branch
National Cancer Institute
National Institutes of Health
Bethesda, MD 20892

INTRODUCTION

Immune intervention began almost two centuries ago when Jenner introduced vaccination with cowpox as a means of protecting against smallpox. This form of immune intervention plays a dominant role in the prevention of human disease. Furthermore, immunological approaches including radioimmunoassays, enzyme-linked immunoassays, microfluorometry, and modern molecular immunogenetics are critical in clinical diagnosis.

Immune intervention is also playing an increasing role in therapy. However, we have not achieved the therapeutic potential provided by the great specificity of the immune system. Most of the medical therapy for cancer still focuses on chemotherapeutic agents acting within the cell. The hybridoma technique of Köhler and Milstein[1] rekindled interest in the use of antibodies targeted to the cell surface as agents to treat cancer patients. However, such monoclonal antibodies have, to date, been relatively ineffective, with only 23 partial and 3 complete remissions reported in the initial 185 patients studied among 25 clinical trials.[2] There have been a number of explanations for this observed low therapeutic efficacy. One of the factors is that the murine monoclonal antibodies are immunogenic. An even more critical factor is that most of the monoclonal antibodies employed are neither cytocidal nor cytostatic agents against human neoplastic cells. Furthermore, in most cases, the antibodies are not directed against a vital structure present on the surface of malignant cells, such as a growth factor receptor required for tumor cell proliferation.

In our laboratory we are readdressing this issue by (a) using genetic engineering to create less immunogenic and more effective monoclonal antibodies, (b) arming antibodies with toxins or radionuclides to enhance their effector functions, and (c) addressing the interleukin-2 receptor (IL-2R) on abnormal cells as a target for effective monoclonal antibody action.[3,4] The scientific basis for this approach is that resting normal cells do

Mechanisms of Lymphocyte Activation and Immune Regulation IV: Cellular Communications
Edited by S. Gupta and T.A. Waldmann, Plenum Press, New York, 1992

57

not express the IL-2R. Rather, this receptor is expressed by a proportion of the activated cells in certain forms of lymphoid neoplasia, in select autoimmune diseases, and in allograft rejection.[3,4] As we shall discuss, therapeutic trials have been initiated to exploit this difference in IL-2R expression, using an unmodified murine monoclonal antibody to the IL-2R, termed anti-Tac. A disadvantage in the use of murine monoclonals is that they are highly immunogenic. In an effort to reduce the immunogenicity of the mouse monoclonal antibodies, monoclonal antibody-mediated therapy has been revolutionized by generating humanized antibodies, including humanized anti-Tac, produced by genetic engineering in which the molecule is human except for the antigen-combining regions, which are retained from the mouse.[5-7] Further, to increase its cytotoxic effectiveness, the anti-Tac monoclonal antibody has been armed with toxins or radionuclides.[8-11] Thus, IL-2 receptor-directed therapy provides a new method for treating certain neoplastic diseases and autoimmune disorders and for preventing allograft rejection.

STRUCTURE AND FUNCTION OF THE MULTISUBUNIT INTERLEUKIN-2 RECEPTOR

To function as effector cells, T cells must change from a resting to an activated state. The sequence of events involved in the activation of T cells begins when a foreign pathogen encounters the antigen-specific receptor on the surface of resting T cells. This antigen-stimulated activation of these resting T cells induces the synthesis of the 15-kDa lymphokine IL-2. To exert its biologic effect, IL-2 must interact with specific high-affinity membrane receptors.[3,4,12-15] Resting cells do not express high-affinity IL-2R, but they are rapidly expressed on T cells after activation with antigen or mitogen.[4,15]

There are three forms of cellular receptors for IL-2: one with a very high affinity (10^{11}/M), one with an intermediate affinity (10^9/M), and another with a much lower affinity (10^8/M) for IL-2. We have used monoclonal antibodies and radiolabeled IL-2 in cross-linking studies to characterize chemically the multiple subunits of this receptor. Initially, a monoclonal antibody (anti-Tac) that reacts with the interleukin-2 binding site of a 55-kDa IL-2R protein (now termed IL-2R α) was identified.[16,17] The receptor protein identified by anti-Tac was characterized as a glycoprotein with an apparent molecular mass of 55 kDa. Using cross-linking methods, we defined a second 70- to 75-kDa (p75 or IL-2R β) IL-2-binding protein.[18,19] We proposed a multisubunit model for the high-affinity IL-2R in which both IL-2R α and IL-2R β proteins are associated in a receptor complex.[18,19] In an independent study, Sharon and colleagues proposed a similar model.[20]

The IL-2R β subunit is a member of a new family of growth and differentiation factor receptors, the hematopoietin receptor superfamily, which includes receptors for IL-2, IL-3, IL-4, IL-6, IL-7, granulocyte-macrophage colony-stimulating factor (GM-CSF), granulocyte colony-stimulating factor (G-CSF), prolactin, growth hormone, and erythropoietin.[21,22] The shared features of the receptors in the hematopoietin receptor superfamily include four conserved cysteine residues located in the N-terminal half of the extracellular ligand-binding domain and a Trp-Ser-X-Trp-Ser motif (WSXWS) located just outside the membrane-spanning domain.

Evidence suggests a more complex subunit structure that involves peptides in addition to the p55 and p75 IL-2-binding peptides. With the use of coprecipitation analysis, radiolabeled IL-2 cross-linking procedures, flow cytometric resonance energy transfer measurements, as well as recovery from photobleaching techniques, a series of peptides of molecular weight 35 000, 40 000, 56 000 (non-IL-2-binding), 75 000 (non-IL-2-binding), 95 000-105 000, 135 000, and 180 000 have been associated with the two IL-2-binding peptides.[23-26] For example, using flow cytometric resonance transfer and lateral diffusion measurements and radiolabeled IL-2 cross-linking studies, we have demonstrated the association of intercellular adhesion molecule 1 (ICAM-1) with the multichain inter-leukin-2 receptor.[23,25,26]

The association of ICAM-1 with the IL-2 receptor may facilitate the paracrine IL-2-mediated stimulation of T cells expressing IL-2 receptors by augmenting homotypic T-T-cell interaction, by receptor-directed focusing of IL-2 release by helper T cells, and by focusing IL-2 receptors of the physically linked cells to the site of lymphocyte function-associated antigen ICAM-1–IL-2 receptor interaction.

INTERLEUKIN-2 RECEPTOR EXPRESSION IN MALIGNANCY OR AUTOIMMUNE DISORDERS

Resting T cells, B cells, or monocytes in the circulation do not display the IL-2 α receptor chain. However, most T and B lymphocytes can be induced to express this receptor subunit. Further, Rubin and coworkers[27] showed that activated normal periph-eral-blood mononuclear cells and certain lines of T- or B-cell origin release a soluble form of the IL-2R α into the culture medium and that normal individuals have measurable amounts of IL-2R α in their plasma. The determination of plasma levels of such IL-2R α provides a valuable noninvasive approach for analyzing both normal and disease-associated lymphocyte activation *in vivo*.

In contrast to the lack of IL-2R α chain expression in normal resting mononuclear cells, this receptor peptide is expressed by a proportion of the abnormal cells in certain forms of lymphoid neoplasia, in select autoimmune diseases, and in allograft rejections. That is, a proportion of the abnormal cells in these diseases expresses surface IL-2R α peptide. Further, the serum concentration of the soluble form of the Tac peptide is elevated in the plasma of such individuals.[3,27] In terms of neoplasia, certain T-cell, B-cell, monocytic, and even granulocytic leukemias express the IL-2R α chain. Specifically, virtually all of the patients with human T-cell lymphotrophic virus-I (HTLV-I)-associated adult T-cell leukemia constitutively express very large numbers of IL-2R α.[28,29] Similarly, a proportion of patients with cutaneous T-cell lymphomas expresses the Tac peptide.[28,30] Further, the malignant B cells of virtually all patients with hairy cell leukemia and a proportion of patients with large- and mixed-cell diffuse lymphomas express IL-2R α.[31] The IL-2R α is also expressed on the Reed-Sternberg cells of patients with Hodgkin's disease and on the malignant cells of patients with true histiocytic lymphoma.[30] Finally, a proportion of the leukemic cells of patients with chronic and acute myelogenous leukemia express the Tac antigen (IL-2R α).

Autoimmune diseases may also be associated with disorders of Tac antigen expres-sion.[32] A proportion of the mononuclear cells in the involved tissues expresses the IL-2R α chain, and the serum concentration of the soluble form of this chain is elevated. Such evidence for T-cell activation and disorders of Tac antigen expression appears in more than 15 autoimmune diseases, including rheumatoid arthritis, systemic lupus erythema-tosus, scleroderma, pulmonary sarcoidosis, and HTLV-1-associated tropical spastic para-paresis. Finally, the Tac peptide is also expressed by the activated lymphocytes in recipients of renal, hepatic, and cardiac allografts that are reacting to the foreign histo-compatibility antigens expressed on the donor organs.[3,32]

DISORDERS OF INTERLEUKIN-2 RECEPTOR EXPRESSION IN HTLV-1-ASSOCIATED ADULT T-CELL LEUKEMIA

A distinct form of mature T-cell leukemia defined by Uchiyama and coworkers was termed adult T-cell leukemia.[33] The retrovirus HTLV-1 was shown to be the primary causative agent in this leukemia.[34] Adult T-cell leukemia is a malignant proliferation of mature CD3/CD4-expressing T cells that infiltrate the skin, lungs, and liver. Cases of adult T-cell leukemia are associated with hypercalcemia and an immunodeficiency state and usually have a very aggressive course with a mean time to death of 20 weeks. The leukemic cells that we and others have examined from patients with HTLV-1-associated adult T-cell leukemia express high- and low-affinity IL-2R, including the IL-2R α chain.[28,29] An analysis of HTLV-1 and its protein products suggests a potential mechanism for this association between HTLV-1 and the constitutive IL-2R α expression. The retrovirus HTLV-1 encodes a 42-kDa protein, now termed *tax*, that is essential for viral replication.[35,36] The *tax* protein, encoded by this retrovirus, also plays a central role in indirectly increasing the transcription of host genes, including the IL-2 and especially the IL-2R α receptor genes involved in T-cell activation and HTLV-1-mediated leukemogenesis.

INTERLEUKIN-2 RECEPTOR α AS A TARGET FOR THERAPY IN PATIENTS WITH HTLV-1-ASSOCIATED ADULT T-CELL LEUKEMIA

Unmodified Anti-Tac Monoclonal Antibody

The HTLV-1-induced adult T-cell leukemia cells constitutively express the IL-2R α chain identified by the anti-Tac monoclonal antibody, whereas normal resting cells do not. This observation provided the scientific basis for IL-2R-directed immunotherapy with this monoclonal antibody. Interleukin-2 receptor-directed immunotherapeutic agents could theoretically eliminate IL-2R α-expressing leukemic cells or abnormally activated T cells involved in other disease states while retaining the Tac-nonexpressing normal T cells and their precursors that express the antigen receptors for T-cell immune responses. In our initial studies, we administered unmodified murine anti-Tac to patients with adult T-cell leukemia.[37,38] The leukemic cells of each patient with adult T-cell leukemia reacted with anti-Tac. Our goal was to inhibit the interaction of IL-2 with its growth factor receptor expressed on the malignant cells. The 19 patients treated in this study did not have untoward reactions related to the immunotherapy and did not have a reduction in the normal formed elements of the blood. Only patients undergoing a remission produced antibodies to the monoclonal antibody. Seven of the 19 treated patients had transient mixed (1), partial (3), or complete remissions (3), lasting from 1 to more than 28 months after anti-Tac therapy. This was assessed by elimination of measurable skin and lymph nodal disease, normalization of serum calcium levels, and routine hematologic and phenotypic tests of circulating cells. Further, elimination of clonal malignant cells was shown by molecular genetic analysis of HTLV-1 proviral integration and the T-cell antigen receptor gene rearrangements. Thus, the use of a monoclonal antibody that prevents the interaction of IL-2 with its growth factor receptor on adult T-cell leukemia cells provides a rational approach for treating this malignancy. Indeed, Maeda and coworkers[39] have presented evidence for the IL-2-dependent expansion of leukemic cells in adult T-cell leukemia in approximately 20% of cases. In most cases of the aggressive phase of adult T-cell leukemia, however, the leukemic cells no longer produce IL-2 nor do they require IL-2 for their proliferation. In this phase of the disease, the patients may not be responsive to unmodified anti-Tac therapy. Nevertheless, such cells continue to display the Tac protein.

Thus, there is still a difference between the normal cells and the malignant cells that can be exploited in treatment. To continue to take advantage of this difference and to improve the effectiveness of IL-2R-directed therapy, the antibody has been armed with cytotoxic agents. For example, *Pseudomonas* exotoxin has been coupled to anti-Tac to deliver a cytotoxic substance directly to the target cancer cells and abnormal T cells.

Before leaving the discussion of unmodified murine anti-Tac, I would like to address the use of this antibody in nonmalignant states, specifically in the prevention of allograft rejection. In transplantation, the foreign transplantation antigens are recognized by the T cells of the host, which then become activated and display the IL-2 receptor. Such Tac-expressing cells participate in allograft rejection. IL-2 receptor-directed antibodies prevent the mixed leukocyte reaction and the resultant generation of specific cytotoxic cells.[17,40] In a study of cardiac transplants in mice, Kirkman et al.[41] administered an antibody directed against the IL-2 receptor to some animals in a transplant model and not to others. Such therapy alone caused a prolongation of graft survival. We obtained similar encouraging results using anti-Tac to prevent allograft rejection in cynomolgus monkeys receiving renal or cardiac allografts.[42,43] With these encouraging results, anti-Tac was used in a phase III trial in which 80 patients receiving renal allografts were entered.[44] One-half of the patients received conventional immunosuppression while the remainder received this immunosuppression supplemented with anti-Tac. Equivalent toxicity has been observed in the two groups. The incidence of early rejection episodes was significantly reduced in cases where anti-Tac was included in the protocol. However, in contrast to the situation with adult T-cell leukemia, these patients made antibodies to the mouse monoclonal antibody.

HUMANIZED ANTIBODY TO THE IL-2 RECEPTOR

There are two problems with murine monoclonal antibodies in general: their immunogenicity and their ineffectiveness at recruiting host-effector functions. We have addressed these issues by producing "humanized" antibodies. These humanized anti-Tac molecules, produced in conjunction with Cary Queen, retain the complementarity-determining region from the mouse, but virtually all the remainder of the molecule is derived from human IgG1 kappa. On the basis of computer modeling of the structure of this antibody, murine elements close to the complementarity-determining regions were identified, and those that were believed to be important to maintain the appropriate conformation of this antibody were retained.[6,7]

One primary goal in these studies is to maintain the affinity and functional capacity of the mouse monoclonal antibody. The parent anti-Tac molecule had an affinity of 9×10^9/M, whereas the hyperchimeric "humanized" version had an affinity of 3×10^9/M, still very high.[6] The parent monoclonal and the humanized version showed a comparable inhibition of T-cell proliferation in response to tetanus antigen.[7]

The humanized version of anti-Tac had improved pharmacokinetics when compared to the murine version with an *in vivo* survival that is 2.5-fold longer (terminal $t_{1/2}$, 103 hr vs. 38 hr). In addition, humanized anti-Tac was less immunogenic than murine anti-Tac when administered to cynomolgus monkeys undergoing heterotopic cardiac allografting.[43] Specifically, all monkeys treated with murine anti-Tac developed measurable anti-murine anti-Tac levels by day 15 (mean onset, 11 days). In contrast, none of the animals receiving humanized anti-Tac produced measurable antibodies to this monoclonal antibody before day 33. Furthermore, the antibody titers in the animals receiving humanized anti-Tac were markedly lower than those receiving the murine monoclonal antibody. A final goal of this project is to make an antibody that is a better effector of cell killing than is murine anti-Tac. Therefore, we were greatly encouraged by the observation that although the parent mouse anti-Tac could not function in antibody-dependent cellular cytotoxicity (ADCC) with human mononuclear cells, the hyperchimeric IgG1 anti-Tac manifests ADCC with human mononuclear cells.[7] With this new ADCC activity, it is hoped that there will be a substantial improvement in the performance of the antibody *in vivo* that should translate into an increase in efficacy in therapy of T-cell leukemia/lymphoma and in the prevention of allograft rejection.

To continue to take advantage of the difference in IL-2R expression and to improve the effectiveness of IL-2R-directed therapy, different approaches have been initiated to modify the antibody for clinical purposes. For example, *Pseudomonas* exotoxin (PE) has been coupled to anti-Tac to deliver a cytotoxic substance directly to the target cells without incurring systemic toxicity. Unfortunately, the initial PE anti-Tac conjugate was hepatotoxic in two of the five adult T-cell leukemia patients treated. Following this observation, Hwang and coworkers[45] performed functional analyses on PE deletion mutants to determine which domains of PE were absolutely required for cell killing and which were responsible for the ubiquitous unwanted binding and hepatic toxicity. Domain III of the three-domain PE exotoxin was responsible for ADP-ribosylating activity. Domain II was involved in the translocation events and therefore must also be retained. However, the 26-kDa domain I of the 66-kDa peptide was involved in the unwanted ubiquitous cell binding and could be eliminated. Therefore, domain I was removed, yielding a 40-kDa peptide, PE40, involving the second and third domains, which alone does not bind to cells. PE40 requires a delivery system to yield specific cell killing. A delivery system was engineered on the molecular level so that IL-2 and the PE elements required for toxicity were joined as a fusion protein, IL-2 PE40, thus providing an alternative (lymphokine-mediated) method of delivering PE40 to the surface of Tac-expressing cells.[8]

Another IL-2R-directed toxin fusion protein, anti-Tac Fv-PE40, has been prepared.[9] It is a fusion protein that contains the variable parts of the heavy and light chains of the mouse monoclonal anti-Tac antibody joined by a linker and further joined to PE40. This single-chain molecule is similar to anti-Tac PE40 but lacks the constant domain. It manifests high-affinity binding for Tac-expressing cells when compared with the parent molecule. Furthermore, it is effective in killing Tac-expressing cells, including those involved in the mixed leukocyte reaction, as well as those freshly obtained from patients with adult T-cell leukemia.[46] Thus, a number of therapeutic approaches have been developed using fusion proteins involving a truncated toxin associated with lymphokines or antibody elements. These agents are being evaluated for toxicity and efficacy in a cardiac allograft model in cynomolgus monkeys prior to their use in the treatment of patients with IL-2R-expressing neoplasias.

In parallel studies, we have turned to β- and α-emitting isotopes as alternative cytotoxic agents that could be conjugated to anti-Tac and are effective when bound to the surface of Tac-expressing cells. In these studies, we have used the β-emitting yttrium-90 (^{90}Y) and the α-emitting bismuth-212 (^{212}Bi). Our choice of isotopes is based on the desire to have agents with a short distance of action that will act on the cell in question and on a small number of bystander cells without unwanted toxicity. In one case, we bound the β-emitting ^{90}Y to anti-Tac using chelates that did not permit elution of radiolabeled yttrium from the monoclonal antibody. Monkeys that received xenografts or allografts of cynomolgus hearts showed a marked prolongation of graft survival following administration of ^{90}Y-labeled anti-Tac.[10,11,47] Following preclinical efficacy and toxicity studies, we initiated a dose escalation trial of ^{90}Y anti-Tac for the treatment of patients with HTLV-1-associated Tac-expressing adult T-cell leukemia. At the doses utilized (5 and 10 mCi per patient) no toxicity was observed in five of the six patients examined, whereas modest granulocytopenia and thrombocytopenia were observed in a single patient. Five of the six patients underwent a partial (3) or complete (2) remission that has been sustained for the 3 to 13 months of observation following initiation of ^{90}Y anti-Tac therapy. Thus, anti-Tac armed with a radionuclide provides meaningful therapy for a form of leukemia that was previously universally fatal. One of the most promising directions for future development of armed monoclonal antibodies for the treatment of cancer involves the linkage of β- and α-emitting radionuclides to human or humanized monoclonal antibodies. Such conjugates may prove to be relatively non-immunogenic agents that are effective in the elimination of IL-2R-α-expressing cells.

In summary, our present understanding of the IL-2/IL-2R system opens the possibility for more specific immune intervention. The IL-2R may prove to be an extraordinarily useful therapeutic target. The clinical applications of anti-IL-2R-directed therapy represent a new perspective for the treatment of certain neoplastic diseases, select autoimmune disorders, and graft-versus-host disease, and for the prevention of allograft rejection.

REFERENCES

1. G. Köhler and C. Milstein, Continuous cultures of fused cells secreting antibody of pre-defined specificity, *Nature* 256:495 (1975).
2. R. Catane and D.L. Longo, Monoclonal antibodies for cancer therapy, *Isr. J. Med. Sci.* 24:471 (1989).
3. T.A. Waldmann, Multichain interleukin-2 receptor: a target for immunotherapy in lymphoma, *J. Natl. Cancer Inst.* 81:914 (1989).
4. T.A. Waldmann, The interleukin-2 receptor, *J. Biol. Chem.* 266:2681 (1991).
5. P.T. Jones, P.H. Dear, J. Foote, M.S. Neuberger, and G. Winter, Replacing the complementarity-determining regions in a human antibody with those from a mouse, *Nature* 321:522 (1986).
6. C. Queen, W.P. Schneider, H.E. Selick, P.W. Payne, N.F. Landolfi, J.F. Duncan, N.M. Avdalovic, M. Levitt, R.P. Junghans, and T.A. Waldmann, A humanized antibody that binds to the IL-2 receptor, *Proc. Natl. Acad. Sci. USA* 86:10029 (1989).
7. R.P. Junghans, T.A. Waldmann, N.D. Landolfi, N.M. Avdalovic, W.P. Schneider, and C. Queen, Anti-Tac-H, a humanized antibody that binds the interleukin-2 receptor: a novel agent for immunotherapy in malignant and immune disorders, *Cancer Res.* 50:1495 (1990).
8. H. Lorberboum-Galski, R. Kozak, T. Waldmann, P. Bailon, D. FitzGerald, and I. Pastan, IL2-PE40 is cytotoxic to cells displaying either the p55 or p75 subunit of the IL-2 receptor, *J. Biol. Chem.* 263:18650 (1988).
9. V.K. Chaudhary, C. Queen, R.P. Junghans, T.A. Waldmann, D.J. FitzGerald, and I. Pastan, A recombinant immunotoxin consisting of two antibody variable domains fused to *Pseudomonas* exotoxin, *Nature* 339:394 (1989).
10. R.W. Kozak, R.W. Atcher, O.A. Gansow, A.M. Friedman, J.J. Hines, and T.A. Waldmann, Bismuth-212 labeled anti-Tac monoclonal antibody: alpha-particle emitting radionuclides as modalities for radioimmunotherapy, *Proc. Natl. Acad. Sci. USA* 83:474 (1986).
11. R.W. Kozak, A. Raubitschek, S. Mirzadeh, M.W. Brechbiel, R. Junghans, O.A. Gansow, and T.A. Waldmann, Nature of the bifunctional chelating agent used for radioimmunotherapy with yttrium-90 monoclonal antibodies: a critical factor in determining *in vivo* survival and organ toxicity, *Cancer Res.* 49:2639 (1989).
12. K.A. Smith, T-cell growth factor, *Immunol. Rev.* 51:337 (1980).
13. T.A. Waldmann, The structure, function, and expression of interleukin-2 receptors on normal and malignant lymphocytes, *Science* 232:727 (1986).
14. W.A. Kuziel and W.C. Greene, Interleukin-2 and the IL-2 receptor: new insights into structure and function, *J. Invest. Dermatol.* 94:27S (1990).
15. R.J. Robb, A. Munck, and K.A. Smith, T cell growth factor receptors. Quantification, specificity, and biological relevance, *J. Exp. Med.* 154:1455 (1981).

16. T. Uchiyama, S. Broder, and T.A. Waldmann, A monoclonal antibody (anti-Tac) reactive with activated and functionally mature human T cells, *J. Immunol.* 126:1393 (1981).

17. T. Uchiyama, D.L. Nelson, T.A. Fleischer, and T.A. Waldmann, A monoclonal antibody (anti-Tac) reactive with activated and functionally mature human T cells, Expression of Tac antigen on activated cytotoxic killer T-cells, suppressor cells, and one of two types of helper T cells, *J. Immunol.* 126:1398 (1981).

18. M. Tsudo, R.W. Kozak, C.K. Goldman, and T.A. Waldmann, Demonstration of a non-Tac peptide that binds interleukin-2: a potential participant in a multichain interleukin 2 receptor complex, *Proc. Natl. Acad. Sci. USA* 83:9694 (1986).

19. M. Tsudo, R.W. Kozak, C.K. Goldman, and T.A. Waldmann, Contribution of a p75 interleukin-2 binding peptide to a high-affinity interleukin-2 receptor complex, *Proc. Natl. Acad. Sci. USA* 84:4215 (1987).

20. M. Sharon, R.D. Klausner, B.R. Cullen, R. Chizzonite, and W.J. Leonard, Novel interleukin-2 receptor subunit detected by cross-linking under high-affinity conditions, *Science* 234:859 (1986).

21. M. Hatakeyama, M. Tsudo, S. Minamoto, T. Kono, T. Doi, T. Miyata, and T. Tomiguchi, Interleukin-2 receptor β chain gene: generation of three receptor forms by cloned human α and β chain cDNA's, *Science* 244:551 (1989).

22. J.F. Bazan, Structural design and molecular evolution of a cytokine receptor superfamily, *Proc. Natl. Acad. Sci. USA* 87:6934 (1990).

23. J. Szöllösi, S. Damjanovich, C.K. Goldman, M.J. Fulwyler, A.A. Aszalos, G. Goldstein, P. Rao, M.A. Talle, and T.A. Waldmann, Flow cytometric resonance energy transfer measurements support the association of a 95-kDa peptide termed T27 with the 55-kDa Tac peptide, *Proc. Natl. Acad. Sci. USA* 84:7246 (1987).

24. M. Sharon, J.R. Gnarra, M. Bamiyash, and W.J. Leonard, Possible association between IL-2 receptors and class I HLA molecules on T cells, *J. Immunol.* 141:3512 (1988).

25. J. Burton, C.K. Goldmann, P. Rao, M. Moos, and T.A. Waldmann, The association of intercellular adhesion molecule 1 with the multichain high-affinity interleukin-2 receptor, *Proc. Natl. Acad. Sci. USA* 87:7329 (1990).

26. M. Edidin, A. Aszalos, S. Damjanovich, and T.A. Waldmann, Lateral diffusion measurements give evidence for association of the Tac peptide of the IL-2 receptor with the T27 peptide in the plasma membrane of HUT-102-B2 T cells, *J. Immunol.* 141:1206 (1988).

27. L.A. Rubin, C.C. Kurman, W.E. Biddison, N.D. Goldman, and D.L. Nelson, A monoclonal antibody 7G7/B6 binds to an epitope on the human interleukin-2 (IL-2) receptor that is distinct from that recognized by IL-2 or anti-Tac, *Hybridoma* 4:91 (1985).

28. T.A. Waldmann, W.C. Greene, P.S. Sarin, C. Saxinger, D.W. Blayney, W.A. Blattner, C.K. Goldman, K. Bongiovanni, S. Sharrow, J.M. Depper, W. Leonard, T. Uchiyama, and R.C. Gallo, Functional and phenotypic comparison of human T cell leukemia/lymphoma virus positive adult T-cell leukemia with human T-cell leukemia/lymphoma virus negative Sézary leukemia, and their distinction using anti-Tac: monoclonal antibody identifying the human receptor for T cell growth factor, *J. Clin. Invest.* 73:1711 (1984).

29. T. Uchiyama, T. Hori, M. Tsudo, Y. Wano, H. Umadome, S. Tamori, J. Yodoi, M. Maeda, H. Sawami, and H. Uchino, Interleukin-2 receptor (Tac antigen) expressed on adult T cell leukemia cells, *J. Clin. Invest.* 76:446 (1985).

30. R. Schwarting, J. Gerdes, and H. Stein, Expression of interleukin-2 receptor on Hodgkin's and non-Hodgkin's lymphoma and macrophages, *J. Clin. Pathol.* 38:1196 (1985).

31. S.J. Korsmeyer, W.C. Greene, J. Cossman, S.M. Hsu, J.P. Jensen, L.M. Neckers, S.L. Marshall, A. Bakhshi, J.M. Depper, W.J. Leonard, E.S. Jaffe, and T.A. Waldmann, Rearrangement and expression of immunoglobulin genes and expression of Tac antigen in hairy cell leukemia, *Proc. Natl. Acad. Sci. USA* 80:4522 (1983).

32. T. Diamantstein and H. Osawa, The interleukin-2 receptor, its physiology and a new approach to a selective immunosuppressive therapy by anti-interleukin-2 receptor monoclonal antibodies, *Immunol. Rev.* 92:5 (1986).

33. T. Uchiyama, J. Yodoi, K. Sagawa, K. Takatsuki, and H. Uchino, Adult T-cell leukemia: clinical and hematologic features of 16 cases, *Blood* 50:481 (1977).

34. B.J. Poiesz, F.W. Ruscetti, A.F. Gazdar, P.A. Bunn, J.D. Minna, and R.C. Gallo, Detection and isolation of type C retrovirus particles from fresh and cultured lymphocytes of a patient with cutaneous T-cell lymphoma, *Proc. Natl. Acad. Sci. USA* 77:7415 (1980).

35. M. Seiki, S. Hattori, Y. Hirayama, and M. Yoshida, Human adult T-cell leukemia virus: complete nucleotide sequence of the provirus genome integrated in leukemia cell DNA, *Proc. Natl. Acad. Sci. USA* 80:3618 (1983).

36. J.G. Sodroski, C.A. Rosen, and W.A. Haseltine, *Trans*-acting transcriptional activation of the long terminal repeat of human T lymphotropic viruses in infected cells, *Science* 225:381 (1984).

37. T.A. Waldmann, C.K. Goldman, K.F. Bongiovanni, S.O. Sharrow, M.P. Davey, K.B. Cease, S.J. Greenberg, and D. Longo, Therapy of patients with human T-cell lymphotrophic virus I-induced adult T-cell leukemia with anti-Tac, a monoclonal antibody to the receptor for interleukin-2, *Blood* 72:1805 (1988).

38. T.A. Waldmann, Monoclonal antibodies in diagnosis and therapy, *Science* 252:1659 (1991).

39. M. Maeda, N. Arima, Y. Daitoku, M. Kashihara, H. Okamoto, T. Uchiyama, K. Shirono, M. Matsuka, T. Hattori, K. Takatsuki, K. Ikuta, A. Shimuzu, T. Honjo, and J. Yodoi, Evidence for the interleukin-2 dependent expansion of leukemic cells in adult T-cell leukemia, *Blood* 70:1407 (1987).

40. J.M. Depper, W.J. Leonard, T.A. Waldmann, and W.C. Greene, Blockade of the interleukin-2 receptor by anti-Tac antibody: inhibition of human lymphocyte activation, *J. Immunol.* 131:690 (1983).

41. R.L. Kirkman, L.V. Barrett, and G.N. Gaulton, Administration of an anti-interleukin-2 receptor monoclonal antibody prolongs cardiac allograft survival in mice, *J. Exp. Med.* 162:358 (1985).

42. M.H. Reed, M.E. Shapiro, T.B. Strom, E.L. Milford, C.B. Carpenter, D.S. Weinberg, K.A. Reimann, N.L. Letvin, T.A. Waldmann, and R.L. Kirkman, Prolongation of primate renal allograft survival by anti-Tac, an anti-human IL-2 receptor monoclonal antibody, *Transplantation* 47:55 (1989).

43. P.S. Brown, Jr., G.L. Parenteau, F.M. Dirbas, R.J. Garsia, C.K. Goldman, M.A. Bukowski, R.P. Junghans, C. Queen, J. Hakimi, W. Benjamin, R.E. Clark, and T.A. Waldmann, Anti-Tac-H, a humanized antibody to the interleukin-2 receptor prolongs primate cardiac allograft survival, *Proc. Natl. Acad. Sci. USA* 88:2663 (1991).

44. R.L. Kirkman, M.E. Shapiro, C.B. Carpenter, D.B. McKay, D.L. Milford, E.L. Ramos, N.L. Tilney, T.A. Waldmann, C.E. Zimmerman, and T.B. Strom, A randomized prospective trial of anti-Tac monoclonal antibody in human renal transplantation, *Transplantation* 51:107 (1991).

45. J. Hwang, D.J. FitzGerald, P. Adhya, and I. Pastan, Functional domains of *Pseudomonas* exotoxin identified by deletion analysis of the gene expressed in *E. coli*, *Cell* 48:129 (1987).

46. R.J. Kreitman, V.K. Chaudhary, T.A. Waldmann, M.C. Willingham, D.J. FitzGerald, and I. Pastan, The recombinant immunotoxin anti-Tac (Fv)-PE40 is cytotoxic toward peripheral blood malignant cells from patients with adult T-cell leukemia, *Proc. Natl. Acad. Sci. USA* 87:8291 (1990).

47. M.M. Cooper, R.C. Robbins, C.K. Goldman, S. Mirzadeh, C. Stone, O.A. Gansow, R.E. Clark, and T.A. Waldmann, Use of yttrium-90-labeled anti-Tac antibody in primate xenograft transplantation, *Transplantation* 50:760 (1990).

LYMPHOCYTE DEVELOPMENT IN MICE DEFICIENT FOR MHC CLASS I EXPRESSION

David H. Raulet, Nan-Shih Liao, Isabel Correa, and Mark Bix

Department of Molecular and Cell Biology
Division of Immunology, 489 LSA
University of California
Berkeley, CA 94720

INTRODUCTION

The role of MHC Class I molecules in recognition of antigens by CD8$^+$ T cells is well established, as is their role in the development of CD8$^+$ T cells. The study of animals unable to normally express class I molecules provides an approach to learn about the role of class I molecules not only in CD8$^+$ T cell development, but possibly in other cell types as well. This can be accomplished by analysis of mice mutant for the light chain of class I MHC, β2-microglobulin, which is necessary for normal functional cell surface expression of class I molecules. Such mice were produced in two laboratories by substitution of the normal β2-microglobulin gene for a mutant one, by homologous recombination embryonic stem cells, which were allowed to repopulate the germ line in chimeric mice [1, 2, 3, 4]. In previous studies, we and others reported that mice homozygous for the mutant β2-m gene have severely diminished cell surface expression of MHC-I molecules, and are severely deficient in the production of functional, mature CD8$^+$CD4$^-$ T cells [2, 4]. In this article, we summarize our recent work on the role of MHC-I molecules in development of CD8$^+$ T cells, CD4$^-$CD8$^-$αβ$^+$ T cells, and γδ$^+$ T cells. In addition the role of MHC-I molecules in the development of natural killer (NK) cells is discussed.

MHC-I AND THE DEVELOPMENT OF MATURE CD8+ T CELLS

In mice homozygous for the mutant β2-m allele (hereafter denoted "-/- mice") the frequencies of CD8+CD4-$\alpha\beta$TCR+ cells in the thymus, spleen and lymph nodes were reduced by at least 20-fold compared to +/- and +/+ (ie wild-type) animals [2]. No difference was evident in the frequencies of CD4+CD8+, CD4+CD8-, or CD8+$\alpha\beta$TCR- cells in the thymus.

We wished to determine the cell types in which MHC-I expression is important for the development of mature, CD8+ T cells. Therefore, we produced irradiation chimeric mice in which fetal liver cells from -/- mice are transferred to lethally irradiated +/+ mice, and a reciprocal set of chimeras in which fetal liver cells from +/+ mice is transferred to lethally irradiated -/- mice. After allowing several weeks for T cells to develop, we could determine whether MHC-I expression in the hematopoietic compartment versus the non-hematopoietic compartment is important for differentiation of CD8+ T cells.

In -/- $\rightarrow$ +/+ fetal liver chimeras, the hematopoietic cells within the thymus are largely of -/- origin, whereas the thymic epithelial cells are of +/+ origin. We found that approximately normal numbers of mature CD8+ T cells develop within these thymi, and can also be found in the periphery, where they are able to mount strong CTL responses to allogeneic cells. These CD8+ T cells are of mutant origin, as shown by the lack of MHC-I on their surfaces. These results indicate that in a thymic environment where thymic epithelial cells are MHC-I+, -/- progenitors efficiently differentiate into functional CD8+ T cells.

In +/+ $\rightarrow$ -/- fetal liver chimeras, the hematopoietic cells within the thymus are largely of +/+ origin, whereas the thymic epithelial cells are exclusively of -/- origin. We found reduced but significant levels of differentiation of CD8+ mature T cells in the thymi of +/+ $\rightarrow$ -/- fetal liver chimeras. As a control, we examined -/- $\rightarrow$ -/- fetal liver chimeras; in those animals the level of differentiation of mature CD8+ T cells was low and similar to that in normal -/- mice. At the functional level, T cells from +/+ $\rightarrow$ -/- chimeras develop strong allo-specific CTL responses, while -/- $\rightarrow$ -/- chimeras fail to develop detectable allospecific CTL. Since the thymic epithelial cells are exclusively of -/- origin in +/+ $\rightarrow$ -/- chimeras, these results suggest that MHC-I+ hematopoietic cells can direct significant development of CD8+ T cells.

The finding that hematopoietic cells may direct CD8+ T cell differentiation, albeit inefficiently, bears on the mechanism of positive selection. Specifically, the results argue against models in which thymic epithelial cells possess unique features necessary to direct T cell development.

MHC-I AND THE DEVELOPMENT OF NATURAL KILLER CELLS

The recognition specificity of natural killer (NK) cells is poorly understood. An inverse relationship between MHC-I expression by tumor target cells and susceptibility to lysis by NK cells has been shown by several groups [13, 14]. We found that even MHC-I-deficient non-tumor cells, that is Con A blast or LPS blast target cells from -/- mice, are lysed efficiently by NK cells [15, 16]. These results indicate that MHC-I deficiency is sufficient to render a normal cell susceptible to NK cells. The mechanism of the "recognition" by NK cells of the absence of MHC-I molecules is not known, although two models are often considered. An NK cell receptor may recognize MHC-I molecules with the result being inhibition of lysis; the absence of MHC-I expression would then lead to lysis of the targets. Alternatively, the absence of MHC-I molecules may unmask a ligand for NK cells on the target cell. Some NK target cells have normal levels of MHC-I suggesting that the lysis of MHC-I-deficient cells may reflect only one of several components of NK activity.

Regardless of the mechanism of recognition by NK cells, the role of MHC-I in the process raises the possibility that MHC-I molecules might also play a role in the development of NK cells. We investigated this by examining NK activity in cells from -/- mice[16]. When the YAC-1 target cell was employed, we found a modest and variable reduction in lysis by enriched NK cells from -/- mice compared to lysis by NK cells from +/+ mice. A more profound effect was observed when we examined lysis of -/- target cells. NK cells from the mutant mice failed to lyse the -/- targets, while, as already noted above, NK cells from +/+ mice lyse -/- targets efficiently. These results suggest that the development of a component of NK activity responsible for lysis of -/- cells is dependent on MHC-I expression. MHC-I molecules may be involved in "positive selection" of these NK cells, or, in the absence of MHC-I expression, NK ligands may be unmasked that "tolerize" NK cells. Other possibilities, for example that expression of an NK receptor depends on MHC-I expression, have not been ruled out.

CONCLUSIONS

The results discussed in this article indicate that MHC-I molecules play a role in the development of at least three lymphocyte lineages, CD8+ T cells, CD4-CD8-αβTCR+ T cells, and NK activity. In the case of CD8+ T cells the results suggest that expression of MHC-I by thymic epithelial cells most efficiently directs CD8+ T cell development, but that expression by hematopoietic cells can also direct CD8+ T cell development, albeit weakly. Similar analyses of the cell types that direct development of CD4-CD8-αβTCR+ T cells, and NK activity are ongoing.

CD4$^-$CD8$^-\alpha\beta$TCR$^+$ T cells represent 5-20% of CD4$^-$CD8$^-$ thymocytes. Although they have a mature phenotype (heat stable antigen-negative) and can respond to stimulation of their TCR, their function is unknown [5, 6]. Their late differentiation indicates that they do not represent an intermediate in the development of other T cell subsets. They may carry out a heretofore unknown function, or perhaps they represent a byproduct of the normal T cell development process. An interesting feature of the cells is their Vβ usage: >50% of the cells express Vβ8, raising the possibility that the cells are subject to some form of selection for receptors that include Vβ8 regions [5, 6].

We found that MHC-I molecules play a role in the development of CD4$^-$CD8$^-\alpha\beta$TCR$^+$ T cells. In -/- mice, the proportion of CD4$^-$CD8$^-$ cells that are $\alpha\beta$TCR$^+$ is usually reduced 2-3 fold compared to +/+ mice. More strikingly, the proportion of CD4$^-$CD8$^-$ cells that are <u>Vβ8.2</u>$^+$ is reduced approximately 5 to 10-fold in -/- compared to +/+ mice. These results suggest that MHC-I expression is important for the differentiation of the CD4$^-$CD8$^-$Vβ8$^+$ cells. MHC-I molecules may play a role in positive selection of these cells, or alternatively, these cells may arise from another subset, eg CD8$^+$ T cells, which depend on MHC-I for their differentiation.

MHC-I AND THE DEVELOPMENT OF $\gamma\delta$TCR$^+$ T CELLS

Several instances of $\gamma\delta^+$ T cells that interact with MHC-I or related molecules have been described, leading to the suggestion that $\gamma\delta^+$ T cells generally recognize MHC-I molecules [7, 8, 9]. The possibility that MHC-I molecules might also play a role in the development of $\gamma\delta^+$ T cells has also been suggested, based on findings with mice transgenic for γ and δ TCR genes that confer MHC-I specificity [10, 11].

We have investigated the status of $\gamma\delta^+$ T cells in -/- mice to determine whether these cells require interactions with MHC-I molecules for their maturation [12]. The results indicate that all the $\gamma\delta^+$ T cell populations investigated seem normal in the -/- mice, including the $\gamma\delta$ cells in the thymus, lymph nodes, spleen, epidermal epithelium, and intestinal epithelium. The cellularity, V gene expression, junctional sequences (in the case of epidermal $\gamma\delta^+$ T cells) and responsiveness to anti-TCR stimulation were all investigated and found to be non-deficient in the -/- mice. These results suggest that the bulk of $\gamma\delta^+$ T cells do not require positive selection by MHC-I molecules for their development [12]. It should be noted that the $\beta 2$-m deficiency results in severe or complete functional deficiency of all MHC-I molecules investigated, including the "nonconventional" MHC-I molecules recognized by some $\gamma\delta^+$ T cells [2, 10, 11].

Our studies also show that most γδ⁺ T cells do not depend upon MHC-I expression for their development. This finding contrasts with the evidence that development of T cells expressing transgenic γδ⁺ TCR does depend on MHC-I expression. Further studies will be necessary to fully evaluate the role of MHC-I in recognition by, and development of, different subsets of γδ⁺ T cells.

The use of the MHC-I deficient animals in the study of disparate lymphocyte lineages illustrates the utility of "gene-knockout" mice as a powerful approach to various biological issues. The approach is particularly powerful in the study of the immune system, because in the laboratory the immune system is not essential for survival of the animal. In future, the combination of new applications of gene targeting methods and interbreeding of various mutant animals promises to yield yet more powerful approaches to immunological and other questions.

ACKNOWLEDGEMENTS

We thank our collaborators Maarten Zijlstra and Rudolf Jaenisch, who produced the β2-m-deficient mice, and with whom we collaborated on much of the work discussed herein.

REFERENCES

1. M. Zijlstra, E. Li, F. Sajjadi, S. Subramani and R. Jaenisch. Germ-line transmission of a disrupted β2-microglobulin gene produced by homologous recombination in embryonic stem cells. *Nature* 342, 435 (1989).
2. M. Zijlstra, M. Bix, N.E. Simister, J.M. Loring, D.H. Raulet and R. Jaenisch. β2-Microglobulin deficient mice lack CD4⁻8⁺ cytolytic T cells. *Nature* 344, 742 (1990).
3. B.H. Koller and O. Smithies. Inactivating the β2-microglobulin locus in mouse embryonic stem cells by homologous recombination. *Proc. Natl. Acad. Sci. USA* 86, 8932 (1989).
4. B.H. Koller, P. Marrack, J.W. Kappler and O. Smithies. Normal development of mice deficient in β2M, MHC class I proteins, and CD8⁺ T cells. *Science* 248, 1227 (1990).
5. B.J. Fowlkes, A.M. Kruisbeek, H. Hon-that, M.A. Weston, J.E. Coligan, R.H. Schwartz and D.M. Pardoll. A novel population of T-cell receptor αβ-bearing thymocytes which predominantly expresses a single Vβ gene family. *Nature* 329, 251 (1987).
6. R.C. Budd, G.C. Meischer, R.C. Howe, R.K. Lees, C. Bron and H.R. MacDonald. Developmentally regulated expression of T cell receptor β chain variable domains in immature thymocytes. *J. Exp. Med.* 166, 577 (1987).

7. L.A. Matis, R. Cron and J.A. Bluestone. Major histocompatibility complex linked specificity of γδ receptor bearing T lymphocytes. *Nature* 330, 263 (1987).

8. M. Bonneville, K. Ito, E.G. Krecko, S. Itohara, D. Kappes, I. Ishida, O. Kanagawa, C.A. Janeway Jr., D.B. Murphy and S. Tonegawa. Recognition of a self major histocompatibility complex TL region product by γδ T-cell receptors. *Proc. Natl. Acad. Sci. USA* 86, 5928 (1989).

9. S. Porcelli, M.B. Brenner, J.L. Greenstein, S.P. Balk, C. Terhorst and P.A. Bleicher. Recognition of a cluster differentiation 1 antigen by human CD4⁻CD8⁻ cytolytic T lymphocytes. *Nature* 341, 447 (1989).

10. F. Wells, S.-J. Gahm, S. Hedrick, J. Bluestone, A. Dent and L. Matis. Requirement for positive selection of γδ receptor-bearing T cells. *Science* 253, 903 (1991).

11. P. Pereira, M. Zijlstra, J. McMmaster, J.M. Loring, R. Jaenisch and S. Tonegawa. Blockade of transgenic gamma delta T cell development in beta-2-microglobulin deficient mice. *EMBO J.* 11, 25 (1992).

12. I. Correa, M. Bix, N.-S. Liao, M. Zijlstra, R. Jaenisch and D. Raulet. Most γδ T cells developm normally in β2-microglobulin-deficient mice. *Proc. Natl. Acad. Sci. USA* 89, 653 (1992).

13. K. Karre, H.G. Ljunggren, G. Piontek and R. Kiessling. Selective rejection of H-2-deficient lymphoma variants suggests alternative immune defence strategy. *Nature* 319, 675 (1986).

14. W.J. Storkus, D.N. Howell, R.D. Salter, J.R. Dawson and P. Cresswell. NK susceptibility varies inversely with target cell class I HLA antigen expression. *J. Immunol.* 138, 1657 (1987).

15. M. Bix, N.-S. Liao, M. Zijlstra, J. Loring, R. Jaenisch and D. Raulet. Rejection of class I MHC-deficient hemopoietic cells by irradiated MHC-matched mice. *Nature* 349, 329 (1991).

16. N. Liao, M. Bix, M. Zijlstra, R. Jaenisch and D. Raulet. MHC class I deficiency: susceptibility to natural killer (NK) cells and impaired NK activity. *Science* 253, 199 (1991).

GENERATION OF MUTANT MICE LACKING
SURFACE EXPRESSION OF CD4 OR CD8
GENE TARGETING

Tak W. Mak, Amin Rahemtulla,
Marco Schilham, Dow Rhoon Koh,
and Wai Ping Fung-Leung

The Ontario Cancer Institute
Departments of Medical Biophysics
and Immunology
University of Toronto
500 Sherbourne Street
Toronto, Ontario
Canada M4X 1K9

ABSTRACT

With the use of homologous recombination in embryonic stem cells, two new strains of mice lacking surface CD4 or CD8 expression have been generated. It is hoped that they will be useful mouse strains for the study of autoimmune diseases, tissue transplantation rejections and tumour rejections.

INTRODUCTION

T cells recognize antigens in the context of prodcuts of the major histocompatibility complex (MHC) [1]. Although the specificity for this recognition resides mainly the T cell antigen receptor [2-4] a pair of surface molecules, CD4 and CD8, also play essential roles when they interact with their ligands. The CD4 molecule is found mainly on helper T cells and is believed to interact with the class II products from the MHC. The CD8 molecule, on the other hand, is found mainly on cytotoxic T cells and is known to interact with MHC class I products [5].

In addition to their functions in the recognition of antigens, these molecules are also thought to be essential for thymic development of T lymphocytes. It is known, for example, that both the positive and negative selections of T lymphocytes occur at a stage in thymic development in which both CD4 and CD8 (double positive) are expressed [6-8]. In addition, it has been demonstrated that the CD4 molecule can be found on precursor cells that are believed to be progenitors of the hemopoietic or lymphopoietic lineages [9]. Finally, the CD8 molecule has also been demonstrated to be expressed on

Mechanisms of Lymphocyte Activation and Immune Regulation IV: Cellular Communications
Edited by S. Gupta and T.A. Waldmann, Plenum Press, New York, 1992

73

a population of earlier thymocytes, before these cells differentiate into double positive cells [10,11] (Figure 1). However, although their expression has been clearly demonstrated, their importance on each of these subsets is not known. For example, (1) is CD4 or CD8 required for the development of all T lymphocytes? (2) is CD4 needed for maturation of CD8 T cells? or (3) is CD8 required for the differentiation of all CD4 T lymphocytes?

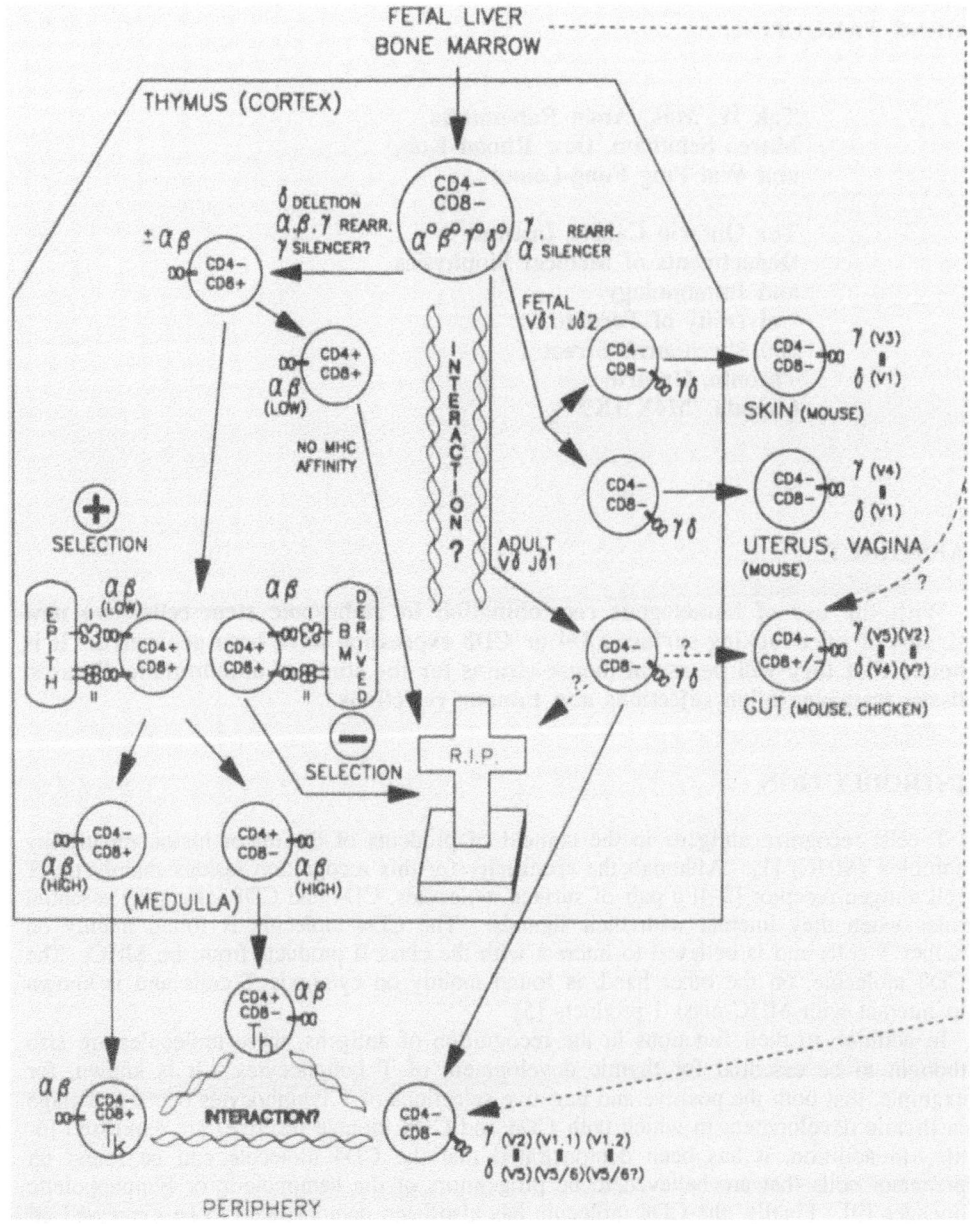

Figure 1. Schematic diagram showing thymic development of T lymphocytes. The TcRαβ thymocytes express CD4 and/or CD8.

While it is clear that many, if not all CD4 T lymphocytes, are helper cells, and those expressing CD8 are cytotoxic T cells, it is not known to what extent can CD8 cells performed the role of helper T cell (in a B cell response or in a cytotoxic response). It is also not certain, what role, if any, CD4 T lymphocytes play in an in vivo killing response.

In order to obtain more definitive answers to the above questions concerning the role of CD4 and CD8 in T cell development and function, one can generate mutant mice that lack cell-surface expression of either the CD4 or CD8 molecules. Such mutant mice can also be used to disclose the importance of each of these subsets of T cells in disease situations. This can be achieved by creating mouse strains with the appropriate genetic backgrounds that are suitable for studying autoimmune diseases, transplantation rejection or tumour rejection. While the creation of mutant mice that lack expression of either CD4 or CD8 hay have been difficult to conceive a few years ago, such goals are now attainable with the use of the technology of homologous recombination in embyronic stem cells (ES).

Using this technology, several groups have pioneered these techniques to obtain mutant mice with specifically targeted mutations [12,13]. ES cells carrying specific mutations can be obtained by electroporation of genomic DNA constructs with specific genetic modifications to generate either specific mutations or interruptions in the exon of the genes concerned. By using the technique of polymerase chain reaction (PCR), rare clones that have undergone homologous recombination with the transfected DNA constructs can be detected. Such ES cells with targeted mutations can then be re-injected into fresh blastocytes for the construction of chimeric mice which would have genetic contributions from these ES lines. These chimeric mice can consequently be bred to obtain germ-line transmissions of the genetic determinants from the gene targeted ES cells. Thus, the targeted mutations generated in the initial manipulations of the embryonic stem cells also have finite chances of being incorporated into the germ-line. By this technology, mutant mice with specific mutations can be created.

GENERATION OF A CD8⁻ MOUSE STRAIN

Using the technology homologous recombination, we have recently generated a mouse strain that does not express surface CD8 [14]. Mice homozygous with respect to the mutation do not express cell surface Lyt-2. This was achieved by disrupting the Lyt-2 gene in the first exon of the gene (by inserting a neomycin resisting gene). The mice with germ-line transmission was bred to homozygosity with respect to the mutant gene. It was found that in the absence a functional Lyt-2 gene, no CD8 was expressed on the surface of the T cells. These data clearly demonstrate that Lyt-2 gene products are required for expression of Lyt-2 / Lyt-3 heterodimers. The results also confirmed that Lyt-3 homodimers cannot emerge in the absence of Lyt-2 gene products, since no surface Lyt-3 can be detected.

Despite the lack of expression of CD8, maturation of CD4 T lymphocytes appears to be unaffected. Normally, levels of CD4 cells can be found both in the thymus and in the peripheral lymph node organs. Functional assays demonstrated that class I restricted cytotoxic T cells are either drastically reduced or absent in these mice. However, the CD4⁺ T lymphocytes appear to be able to respond well to class II restricted targets. Furthermore, functional assays and lymphokine productions in response to allo-or viral antigens can be readily demonstrated in these CD8⁻ mice. These results indicated that this new mouse strain has an immune system that possesses a normal CD4 helper compartment but lacks MHC class I restricted cytotoxic T cells [14].

GENERATION OF A MOUSE STRAIN DEVOID OF CD4 CELLS

Using homologous recombination in ES cells, we have also recently generated a mouse strain that does not express surface CD4. This was achieved by interrupting the fifth exon of the L3T4 gene by the insertion of sequences of the neomycin resistance gene in these coding sequences. Studies indicated that no CD4$^+$ cells can be found in these mutant mice. However, there are normal levels of CD8$^+$ thymocytes and T lymphocytes. These results indicate that mature T lymphocytes can develop in the absence of CD4 suggesting that surface expression of these molecules is not required for differentiation of all T cells. Thus, its expression on hemopoietic and lymphopoietic precursor cells as well as on the CD4/CD8 double positive thymocytes is not absolutely necessary for the development of all T cells, for example, CD8$^+$ $\alpha\beta^+$ lymphocytes. Functional analyses of these mice also indicate that they can mount successful cytotoxic responses against viruses (15). T cell "help", however, was reduced in these mice without CD4 cells.

THE ROLE OF A CD4 OR CD8 CELLS IN DISEASE SITUATIONS

During the last 10 years, many investigators have attempted to dissect the roles of CD4 helper T cells and CD8 killer T cells in a variety of disease situations in which immune cells are implicated. These include systems in which animal models are available. For example, autoimmune-like diseases can develop in certain inbred strains of mice. Diabetes is known to develop in (NOD) mice, while lupus-like diseases can be found in the mouse strains of MRL/1pr and NZB backgrounds. Similarly, experimental autoimmune encephalomyelitis can be induced in mice with haplotype H-2^u, by injection of myelin basic protein, and arthritis-like disease can develop in certain strains of mice (like those with H-2^q hapotype) after injection of collagen. Similarly, transplant rejection in animal models, as well as the rejction of tumours in transplanted mice, can also be examined. The availability of these CD4$^-$ and CD8$^-$ mice opens up the possibility of directly testing the roles of each of the T cell subsets in the disease processes in these experimental model systems. It is hoped that such studies, with these mutant strain of mice, will help us further to understand the process of immune destruction or rejections in these animal systems and, consequently, those in man.

REFERENCES

1. R.H. Schwartz, T-lymphocyte recognition of antigen in assocation with gene products of the major histocompatibility complex, *Ann. Rev. Immunol.* 3:237 (1985).
2. S.M. Hedrick, D.I. Cohen, E.A. Nielsen, and M.M. Davis, Isolation of cDNA clones encoding T cell specific membrane associated proteins, *Nature* 308:149 (1984).
3. Y. Yanagi, Y. Yoshikai, K. Leggeth, S.P. Clark, I. Aleksander, and T.W. Mak. A human T cell specific cDNA clone encodes a protein having extensive homology to immunoglobulin chains, *Nature* 308:145 (1984).
4. Toyonaga, B. and T.W. Mak, Genes of the T cell antigen receptor in normal and malignant T cells, *Ann. Rev. Immunol.* 5:585 (1987).
5. J.R. Parnes, Molecular biology and function of CD4 and CD8, *Adv. Immunol.* 44:265 (1989).
6. D.A. Ferrick, P.S. Ohashi, V. Wallace, M. Schilham, and T.W. Mak, Thymic ontogeny and selection of $\alpha\beta$ and $\gamma\delta$ T cells, *Immunol. Today* 10:403 (1989).
7. H. Boehmer, Tolerance in T cell receptor transgenic mice involves deletion of nonmature CD4$^+$8$^+$ thymocytes, *Nature* 333:742 (1988).
8. W.C. Sha, C.A. Nelson, R.D. Newberry, D.M. Kranz, J.H. Russell, and D.Y. Loh, Positive and negative selection of an antigen receptor on T cells in transgenic mice, *Nature* 336:73 (1988).

9.   L. Wu, R. Scollay, M. Egerton, M. Pearse, G.J. Sprangrude, and K. Shortman, CD4 expressed on earliest T-lineage precursor cells in the adult murine thymus, *Nature* 349:71 (1991).

10.   C.J. Guidos, I.L. Weissman, and B. Adkins, Intrathymic maturation of murine T lymphocytes from $CD8^+$ precursors, *Proc. Natl. Acad. Sci. USA* 86:7542 (1989).

11.   H.R. MacDonald, R.C. Rudd, and R.C. Howe, A $CD3^-$ subset of $CD4^-8^+$ thymocytes: a rapidly cycling intermediate in the generation of $CD4^+8^+$ cells, *Eur. J. Immunol.* 18:519 (1988a).

12.   O. Smithies, R.G. Gregg, S.S. Boggs, M.A. Koralewski, and R.S. Kuchenapati, Insertion of DNA sequences into the human chromosomal β-globin locus by homologous recombination, *Nature* 317:230 (1985).

13.   K.R. Thomas and M.R. Capecchi, Site-directed mutagenesis by gene targeting in mouse embryo-derived stem cells, *Cell* 51:503 (1987).

14.   W.-P. Fung-Leung, M.W. Schilham, A. Rahemtulla, T.M. Kündig, M. Vollenweider, J. Potter, W. van Ewijk, and T.W. Mak, CD8 is needed for development of cytotoxic T cells but not helper T cells, *Cell* 65:443 (1991).

15.   A. Rahemtulla, W.-P. Fung-Leung, M.W. Schilham, T.M. Kündig, S.R. Sambhara, A. Narendran, A. Arabian, A. Wakeham, C. Paige, R.M. Zinkernagel, R.G. Miller, and T.W. Mak, Mice lacking CD4 have normal development and function of $CD8^+$ cells but have markedly decreased helper cell activity, *Nature* 353:180 (1991).

ALTERATION OF T CELL LINEAGE COMMITMENT BY EXPRESSION OF A HYBRID CD8/CD4 TRANSGENE

Rho H. Seong and Jane R. Parnes

Department of Medicine
Division of Immunology
Stanford University School of Medicine
Stanford University, Stanford, CA 94305

INTRODUCTION

Thymocyte differentiation and the subsequent production of peripheral functional T cells involve positive and negative selection events on developing thymocytes expressing a number of specific surface molecules, including the T cell receptor (TCR) and the coreceptor molecules CD4 and CD8. At least two different mechanisms may be responsible for the effects of CD4 and CD8 on T cell development. CD4 binding to major histocompatibility complex (MHC) class II proteins[1] and CD8 binding to MHC class I proteins[2] may increase the strength of interaction between the cells involved in the selection processes.[3] Signal transduction through these coreceptors may also play an important role during thymocyte development. Recent evidence indicates that CD4 and CD8 play important roles in signal transduction during T cell activation.[4-9]

Most cells in the $CD4^+CD8^-$ lineage are helper T cells, while most cells in the $CD4^-CD8^+$ lineage are cytotoxic T cells. Both cell types are derived from a common thymic precursor which expresses both CD4 and CD8 ($CD4^+CD8^+\alpha\beta TCR^+$), but it is not clear how these "double positive" thymocytes differentiate to the CD4 or CD8 lineage. Experiments using $\alpha\beta$TCR transgenic mice[10-13] suggest that the CD4/CD8 phenotype of mature T cells is determined by the specificity of the TCR. However, these results do not explain the mechanism of lineage commitment. Two possibilities have been suggested to explain the commitment of each T cell to the CD4 or CD8 lineage: an instructional vs. a stochastic model.[14,15] The stochastic model suggests that developing $CD4^+CD8^+\alpha\beta TCR^+$ thymocytes

Mechanisms of Lymphocyte Activation and Immune Regulation IV: Cellular Communications,
Edited by S. Gupta and T.A. Waldmann, Plenum Press, New York, 1992

79

will randomly lose either CD4 or CD8, and only those cells expressing the appropriate pair of TCR and coreceptor will be further selected. It is also possible that cells expressing an inappropriate pair of TCR and coreceptor (i.e., class I restricted TCR with CD4 or class II restricted TCR with CD8) may be exported to the periphery but not detected in normal antigen responses. The instructional model implicates a specific signal which differs in developing thymocytes. The specific signal depends on whether the αβTCR on the thymocytes binds to either class I or class II thymic MHC molecules. The signal may be delivered through the αβTCR complex, through the αβTCR together with a coreceptor, or through a coreceptor itself, and results in the cell maintaining expression of the appropriate coreceptor and turning off expression of the inappropriate one.

The instructional model has been favored because results from recent experiments[16,17] have not been consistent with the stochastic model. The stochastic model predicts that thymocytes which express a class I restricted transgenic αβTCR but have randomly become CD4$^+$CD8$^-$ will be rescued if a CD8 transgene is expressed constitutively This would produce a new population of mature T cells expressing the class I restricted transgenic αβTCR and CD4 (as well as the constitutive transgenic CD8). However, such cells were not detected in mice expressing a constitutively expressed CD8 transgene and a transgenic class I restricted αβTCR.[15,16] However, premature expression of both the class I restricted transgenic TCR and CD8 may have skewed the results towards not producing CD4$^+$ cells. Furthermore, these results do not show the presence of a signal required to drive cells to a CD4/CD8 cell lineage. We now show that a specific signal for differentiation of an immature thymocyte to the CD4 or CD8 lineage is delivered by each coreceptor molecule through its transmembrane region and/or cytoplasmic tail. We generated transgenic mice (β-AprCD8/4$^+$) expressing a hybrid protein composed of the CD8α extracellular domains linked to the CD4 transmembrane region and cytoplasmic tail. By crossing the β-AprCD8/4$^+$ transgenic mice with mice expressing a transgenic αβTCR specific for a particular antigen plus class I MHC protein, we were able to express the hybrid molecule in developing thymocytes expressing the class I MHC restricted TCR. Our results show that the signal transduced by the hybrid molecule results in the differentiation of immature thymocytes expressing a class I restricted TCR into mature T cells expressing CD4 and are most consistent with an instructional model for T cell differentiation in the thymus.

MATERIALS AND METHODS

mAbs. mAb T3.70 (ref. 18) was kindly provided by Dr.H.-S. Teh. Phycoerythrin (PE)-conjugated anti-CD4, FITC-conjugated anti-CD8α, and allophycocyanin (APC)-conjugated streptavidin were purchased from Becton-Dickinson. mAb 53-5 (ref. 19) is specific for mouse CD8β and culture supernatant of the hybridoma producing this mAb was used to stain cells. Texas red-conjugated mouse anti-rat antibody was purchased from Jackson ImmunoResearch Laboratories, Inc.

Mice. C57BL/6J, C3H/J x DBA/2J F1, and DBA/2J mice were purchased from the Jackson Laboratory. $\alpha\beta TCR^+$ transgenic mice were kindly provided by Dr. H. von Boehmer.

Construction of β-AprCD8/4. The CD8/CD4 cDNA hybrid construct has been described.[7] The encoded protein consists of the signal peptide (amino acids -24 to -1) and extracellular domains (amino acids 1-167) of CD8α (Lyt-2.2) and the transmembrane region and cytoplasmic tail of CD4 (amino acids 374-435). The junctional amino acid sequence for the hybrid protein is FACD/ILA. The hybrid cDNA was placed under the control of human β-actin regulatory sequences.[20] A Bgl II-Eco RI fragment (2.7kb) from the mouse CD4 gene[21] was inserted at the 3' end of the hybrid cDNA to provide a poly(A) signal.

Production of Transgenic Mice. An 8kb SalI-SpeI fragment of the CD8/CD4 hybrid gene construct was isolated and was used for production of transgenic mice by microinjection into C3H/J x DBA/2J F2 embryos (Lyt2.1) as described.[3]

Bone Marrow Transplantation. Bone marrow cells (4×10^6 cells) from female $\alpha\beta TCR^+$, $\alpha\beta TCR^+$ plus β-AL2.1$^+$, or $\alpha\beta TCR^+$ plus β-AprCD8/4$^+$ transgenic mice were transferred into lethally irradiated (900R) C57BL/6J female mice. Three to five weeks after the transfer thymus and lymph nodes from the bone marrow recipients were analyzed.

FACS Analysis of Cells. After staining with mAbs, a total of 20,000 viable cells per sample were analyzed on a modified FACS II using Facscan Research SoftwareProgram (Becton-Dickson).

RESULTS AND DISCUSSION

We predicted that during thymocyte development a CD8/CD4 hybrid protein consisting of CD8α extracellular domains and the transmembrane region and cytoplasmic tail of CD4 would bind to class I MHC proteins but deliver the intracellular signal(s) of CD4. Therefore, when the CD8/CD4 hybrid coreceptor is expressed together with a class I restricted αβTCR on developing thymocytes, the hybrid coreceptor and the αβTCR should bind to the same class I MHC protein expressed on the thymic epithelium. If coreceptor molecules deliver a specific signal for further differentiation of thymocytes to the CD4 or CD8 lineage, then the hybrid coreceptor (by virtue of its CD4 transmembrane and cytoplasmic tail) would be predicted to deliver a signal that will lead thymocytes to the CD4 instead of CD8 lineage. The result would be cells expressing CD4 and the class I restricted transgenic TCR as depicted in Figure 1.

The transgenic mice generated with the hybrid coreceptor construct (β-AprCD8/4$^+$) express the hybrid protein at least in all lymphocytes, including developing thymocytes.[22] In order to express the CD8/CD4 hybrid coreceptor in cells expressing a defined class I restricted αβTCR, the β-AprCD8/4$^+$ transgenic mice (H-2^d) were crossed with mice (H-2^b) expressing a transgenic αβTCR specific for the male (H-Y) antigen in the context of the class I MHC

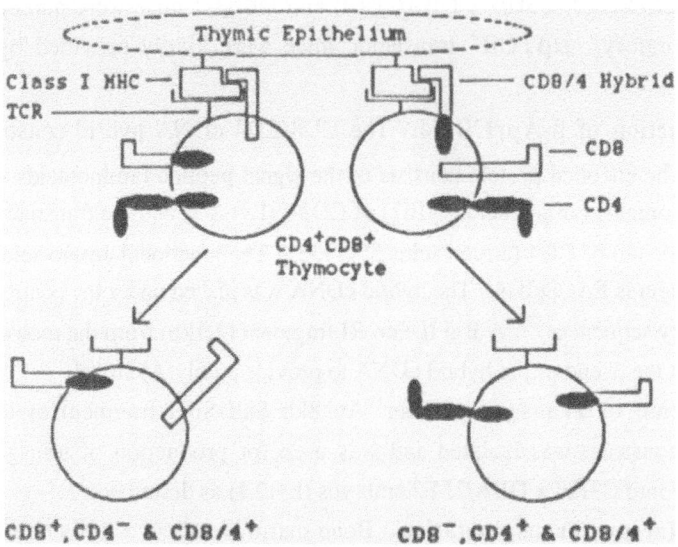

Figure 1. Expression of a transgenic CD8/CD4 hybrid protein on developing thymocytes leads to CD4 lineage commitment. A transgenic CD8/CD4 hybrid protein expressed on a developing thymocyte may bind to the same class I MHC protein as a class I restricted TCR by virtue of its CD8 extracellular domain, but the transmembrane region and/or cytoplasmic tail of CD4 delivers a specific signal which leads the thymocyte to the CD4 lineage to produce a mature $CD8^+CD4^-CD8/4^+$ cell expressing class I restricted TCR. In contrast, binding of native CD8 or a transgenic native CD8α to the same class I MHC protein as the class I restricted TCR leads cells to the CD8 and not the CD4 lineage (see text).

protein H-2D^b (ref. 23). Transgenic αβTCR expression was monitored using the mAb T3.70. This mAb is specific for the transgenic α-chain but stains 10-fold better (T3.70^{hi}) when this α-chain is associated with the transgenic β-chain as opposed to other β-chains.[18] As a result of positive selection, most T cells expressing high levels of both the transgenic TCR α- and β-chain are CD4$^-$CD8$^+$ in αβTCR female transgenic mice of the H-2^b haplotype.[24] The CD4$^+$CD8$^-$ T cells found in these mice result from incomplete allelic exclusion of the TCR α-chain.[18,24] However, in αβTCR female transgenic mice of the H-2^d haplotype, at least some of the CD4$^+$CD8$^-$ T cells also express high levels of the transgenic TCR α-chain.[10] Since the MHC haplotype of the β-AprCD8/4$^+$ transgenic mice is H-2^d, and thymic epithelial cells rather than bone marrow derived cells are responsible for positive selection,[24-28] bone marrow cells from female αβTCR$^+$ transgenic mice or αβTCR plus hybrid CD8/CD4 double transgenic mice were transferred to lethally irradiated female C57BL6/J (H-2^b) mice. The H-2^b background avoids selection of CD4$^+$CD8$^-$ T cells expressing the transgenic TCR α- and β-chains in the bone marrow recipients. In addition, bone marrow cells from female mice expressing both the transgenic αβTCR and a native CD8α transgene (driven by the human β-actin regulatory region) from the previously described β-AL2.1$^+$ transgenic mice[3] were also transferred to X-irradiated female C57BL6/J mice. Analyses of recipients of bone marrow cells showed that in bone marrow recipients from the αβTCR$^+$ transgenic mice and the αβTCR$^+$ plus β-AL2.1$^+$ double transgenic mice, 5-8% of lymph node cells expressing the transgenic TCR$^+$ (T3.70^{hi}) express CD4 (ref. 22). In contrast, in recipients of bone marrow cells from the αβTCR$^+$ plus β-AprCD8/4$^+$ double transgenic mice, about 50% of lymph node cells expressing the transgenic TCR also express

CD4 (ref. 22). Staining of thymocytes also showed an increased proportion of T3.70^{hi}CD4$^+$ cells in the the αβTCR$^+$ plus β-AprCD8/4$^+$ recipients. In the transgenic αβTCR$^+$ or the αβTCR$^+$ plus β-AL2.1$^+$ recipients, only 1-2% of T3.70^{hi} thymocytes express CD4 but not endogenous CD8 (CD4$^+$CD8^{endo-}), which can be identified by staining with mAb 53-5 specific for CD8β (ref. 22). In contrast, in the αβTCR$^+$ plus β-AprCD8/4$^+$ recipients, 4.9% of T3.70^{hi} thymocytes are CD4$^+$CD8^{endo-} (ref. 22). The marked increase in the proportion of T3.70^{hi} CD4$^+$CD8^{endo-} cells in lymph nodes and the thymus from the αβTCR$^+$ plus β-AprCD8/4$^+$ recipients compared to the transgenic αβTCR$^+$ or the αβTCR$^+$ plus β-AL2.1$^+$ recipients, implies that expression of the CD8/CD4 hybrid coreceptor leads immature thymocytes expressing a class I restricted TCR to develop into mature T cells expressing CD4. It is also noted that T3.70^{hi}CD4$^-$CD8^{endo+} cells were markedly reduced in the thymus of αβTCR$^+$ plus β-AprCD8/4$^+$ recipients (8.6%) compared to that of transgenic αβTCR$^+$ or αβTCR$^+$ plus β-AL2.1$^+$ recipients (46.6% and 37.8%), while T3.70^{hi}CD4$^+$CD8^{endo+} cells were found to be twice as abundant in the thymus of αβTCR$^+$ plus β-AprCD8/4$^+$ recipients (56.1%) as compared to transgenic αβTCR$^+$ (27.2%) or αβTCR$^+$ plus β-AL2.1$^+$ recipients (26.2%) (ref. 22). This resulted in a dramatic difference in the CD4$^+$:CD8^{endo+} ratio in the thymus among these bone marrow recipients. The CD4$^+$:CD8^{endo+} ratio in the αβTCR$^+$ plus β-AprCD8/4$^+$ recipients was 1:1.8 (4.9% vs 8.6%), while the ratio in the transgenic αβTCR$^+$ recipients was 1:25 (1.9% vs 46.6%) and that in the αβTCR$^+$ plus β-AL2.1$^+$ recipients was 1:38 (1.0% vs 37.8%). The skewing of the CD4$^+$:CD8^{endo+} ratio in the αβTCR$^+$ transgenic mice as compared to normal mice (2:1) is a result of positive selection of this TCR on CD8^{endo+} cells.[24] The further skewing of this ratio in the αβTCR$^+$ plus β-AL2.1$^+$ recipients is analogous to results described by others in similar transgenics.[16] The "improved" CD4$^+$:CD8^{endo+} ratio in the αβTCR$^+$ plus β-AprCD8/4$^+$ recipients is due to the severe reduction in the T3.70^{hi}CD4$^-$CD8^{endo+} cell population and the increase in the T3.70^{hi} CD4$^+$CD8^{endo-} cell population. The marked reduction in the percentage of the T3.70^{hi}CD4$^-$ CD8^{endo+}cells in the αβTCR$^+$ plus β-AprCD8/4$^+$ recipients (8.6%) compared to that in the αβTCR$^+$ recipients (46.6%) and the αβTCR$^+$ plus β-AL2.1$^+$ recipients (27.8%) (ref. 22) suggests that the T3.70^{hi}CD4$^+$(CD8^{endo+} or CD8^{endo-}) cells in the αβTCR$^+$ plus β-AprCD8/4$^+$ recipients originate from cells that would have otherwise been in the T3.70^{hi}CD4$^-$ CD8^{endo+} population. Thus, the T3.70^{hi}CD4$^+$ cells in the αβTCR$^+$ plus β-AprCD8/4$^+$ recipients seem to be instructed to differentiate to the CD4 lineage instead of becoming T3.70^{hi}CD4$^-$CD8^{endo+} as a result of expression of the transgenic CD8/CD4 hybrid coreceptor. It seems less likely that the T3.70^{hi}CD4$^+$ cells were pre-committed to a CD4 lineage and were then rescued by a signal through the CD4 transmembrane and cytoplasmic domain of the transgenic CD8/CD4 hybrid, because this process would not lead to such a severe reduction in the percentage of T3.70^{hi}CD4$^-$CD8^{endo+} in the αβTCR$^+$ plus β-AprCD8/4$^+$ recipients. Furthermore, as observed by others[16,17] and considered to be inconsistent with a stochastic model for T cell development, the expression of native CD8α in the αβTCR$^+$ plus β-AL2.1$^+$ recipients did not increase the number of T3.70^{hi}CD4$^+$ cells as compared to the the αβTCR$^+$ recipients (5.1% vs 5.96% of T3.70^{hi} cells in lymph nodes; 1.0% vs 1.9% of T3.70^{hi} cells in thymus[22]). These results imply that the CD8/CD4 hybrid protein may bind to the same thymic class I MHC protein as the transgenic $\alpha\beta$TCR (presumably H-2^b) by virtue of its CD8α extracellular domain, but that the transmembrane region and/or cytoplasmic tail of CD4 delivers a specific signal which leads developing thymocytes to the CD4 lineage.

Analyses of lymph node cells from the bone marrow recipients revealed the rather surprising result that 16% of the T3.70^{hi} cells (31% of T3.70^{hi}CD4$^+$ cells) express both CD4 and endogenous CD8 in the $\alpha\beta$TCR$^+$ plus β-AprCD8/4$^+$ recipients.[22] It is unlikely that T3.70^{hi}CD4$^+$CD8^{endo+} cells, which are found only in the $\alpha\beta$TCR$^+$ plus β-AprCD8/4$^+$ recipients, originate from cells that were pre-committed to the CD8 lineage and then rescued to express CD4 by a signal through CD4 transmembrane and cytoplasmic tail of the hybrid coreceptor. If this were the case then one would expect to find a similar population of cells in the periphery of the $\alpha\beta$TCR$^+$ plus β-AL2.1$^+$ recipients (i.e., T3.70^{hi}CD4$^+$CD8^{endo+} cells), because by this logic cells pre-committed to the CD4 lineage in these recipients would have been induced to express CD8^{endo} by a signal through the transgenic native CD8α. Such cells were not detected in these mice.[22] Escape of immature thymocytes from the thymus before full maturation because of the presence of the transgenic hybrid CD8/CD4 molecule could produce the T3.70^{hi} CD4$^+$CD8^{endo+} peripheral T cells in the $\alpha\beta$TCR$^+$ plus β-AprCD8/4$^+$ recipients. However, this also seems unlikely because CD4$^+$CD8^{endo+} peripheral cells were not detected in the TCR$^+$ plus β-AL2.1$^+$ recipients. Since endogenous CD8 is only maintained on 31% of the T3.70^{hi}CD4$^+$ cells in the $\alpha\beta$TCR$^+$ plus β-AprCD8/4$^+$ recipients, it is unlikely that the hybrid CD8/CD4 delivers an incomplete signal which only functions in maintaining the expression of CD4, but not in turning off endogenous CD8. The most likely explanation for the origin of these T3.70^{hi}CD4$^+$CD8^{endo+} cells is that in some cases both the transgenic CD8/CD4 hybrid and endogenous CD8 delivered independent signals into the same cells during thymic selection because of competition between the two molecules for binding to the same class I MHC as the TCR on the selecting cells. Transmission of two different signals in some cells could function to maintain both CD4 and CD8^{endo} expression in those cells. A separate signal might be required to turn off one of the molecules and the "on" signal may be dominant. Alternatively, CD4 and CD8 may be programmed to be turned off during thymocyte maturation, and a signal delivered by the corresponding molecule during positive selection may be required for continued expression of one of these proteins.

These results demonstrate that the CD4 and CD8 coreceptor molecules play an essential role in signalling and directing the developmental pattern of thymocytes, and that a specific signal for differentiation of an immature thymocyte to the CD4 or CD8 lineage is delivered by each molecule, respectively, through its transmembrane region and/or cytoplasmic tail. These results are most consistent with an instructional model for T cell differentiation in the thymus. The tyrosine kinase p56^{lck} is associated with the cytoplasmic tail of CD4 and CD8 (ref. 29-31) and is important for the function of these proteins both in mature and immature T cells. p56^{lck} associated with these coreceptors is likely to play a key role in the signalling process during thymocyte differentiation. However, both CD4 and CD8 associate with p56^{lck}. Therefore, if p56^{lck} is responsible for this signalling, differential signals for commitment to either the CD4 or CD8 lineage may be explained by other molecules which are differentially associated with CD4 versus CD8, by dosage effects due to differences in the level of association between CD4 and CD8 with p56^{lck}, or by differences of accessibility to target molecules by p56^{lck} associated with CD8 versus CD4.

ACKNOWLEDGEMENTS

This work was supported by NIH grants AI 19512 and CA 46507 to JRP and JRP is an Established Invetigator of the American Heart Association.

REFERENCES

1. C. Doyle. and J.L. Strominger, Interaction between CD4 and class II MHC molecules mediates cell adhesion, *Nature* 330:56 (1987).
2. A. Norment, R.D. Salter, P. Parham, V.H. Englehard, and D.L. Littman, Cell-cell adhesion mediated by CD8 and MHC class I molecules, *Nature* 336:79 (1988).
3. R.H. Seong, C.H. Liaw, S. Michie, J.R. Parnes, and J.W. Chamberlain, Effects of high level constitutive expression of transgenic CD8α on thymic selection, Submitted.
4. A. Veillette, A. Brookman, E.M. Horak, L.E. Samelson, and J.B. Bolen, Signal transduction through the CD4 receptor involves the activation of the internal membrane protein kinase p56^{lck}, *Nature* 338:257 (1989).
5. A. Veillette, J.B. Bolen, and M.A. Brookman, Alteration in tyrosine phosphorylation induced by antibody-mediated cross-linking of CD4 receptor of T lymphocytes, *Mol. Cell. Biol.* 9:4441 (1989).
6. K. Luo and B.M. Sefton, Crosslinking of T cell surface molecules CD4 and CD8 stimulates phosphorylation of the lck tyrosine protein kinase at the autophos-phorylation site, *Mol. Cell. Biol.* 10:5305 (1990).
7. R. Zamoyska, P. Durham, S.D. Gorman, P. von Hoegen, J.B. Bolen, A.V. Veillette, and J.R. Parnes, Inability of CD8α' polypeptide to associate with p56^{lck} correlates with impaired function *in vitro* and lack of expression *in vivo*, *Nature* 342:278 (1989).
8. M.C. Miceli, P. von Hoegen, and J.R. Parnes, Coreceptor versus adhesion function of CD4 and CD8: role of the cytoplasmic tail in coreceptor activity, *Proc. Natl. Acad. Sci. USA.* 88:2623 (1991).
9. N. Glaichenhaus, N. Shastri, D.R. Littman, and J.M. Turner, Requirement for association of p56^{lck} with CD4 in antigen-specific signal transduction in T cells, *Cell* 64:511 (1991).
10. H.-S. Teh, P.Kisielow, B. Scott, H. Kishi, H. Uematsu, Y. Bluthmann, and H. von Boehmer, Thymic major histocompatibility complex antigens and the alpha beta T-cell receptor determine the CD4/CD8 phenotype of T cells, *Nature* 335:229 (1988).

11. W.C. Sha, C.A. Nelson, R.D. Newberry, D.M. Kranz, J.H. Russel, and D.Y. Loh, Selective expression of an antigen receptor on CD8-bearing T lymphocytes in transgenic mice, *Nature* 335:271 (1988).

12. L.J. Berg, A.M. Pullen, B. Fazekas de St. Groth, D. Mathis, C. Benoist, and M.M. Davis, Antigen/MHC-specific T cells are preferentially exported from thymus in the presence of their MHC ligand, *Cell* 58:1035 (1989).

13. J. Kaye, M-L Hsu, M.-E. Sauron, S.C. Jameson, N.R. Gascoigne, and S.M. Hedrick, Selective development of CD4$^+$ T cells in transgenic mice expressing class II MHC-restricted antigen receptor, *Nature* 341:746 (1989).

14. H. von Boehmer, P. Kisielow, H. Kishi, B. Scott, P. Borgulya, and H.-S. Teh, The expression of CD4 and CD8 accessory molecules on mature T cells is not random but correlates with the specificity of the ab receptor for antigen, *Immunol. Rev.* 109:144 (1989).

15. E.A. Robey, B.J. Fowlkes, and D.M. Pardoll, Molecular mechanisms for lineage commitment in T cell development, *Semin. Immnol.* 2:25 (1990).

16. E.A. Robey, B.J. Fowlkes, J.W. Gordon, D. Kioussis, H. von Bohmer, F. Ramsdell, and R. Axel, Thymic selection in CD8 transgenic mice supports an instructive model for commitment to a CD4 or CD8 lineage, *Cell* 64:99 (1991).

17. P. Borgulya, H. Kishi, U. Muller, J. Kirberg, and H. von Boehmer, Development of the CD4 and CD8 lineage of T cells: instruction versus selection, *EMBO J.* 10:913 (1991).

18. H.-S. Teh, H. Kishi, B. Scott, and H. von Bohmer, Deletion of autospecific T cells in T cell receptor (TCR) transgenic mice spares cells with normal TCR levels and low levels of CD8 molecules, *J. Exp. Med.* 169:795 (1989).

19. J.A. Ledbetter, W. Seaman, T. Tsu, and L. A. Herzenberg, Ly-2 and Ly-3 antigens are on two different polypeptide subunits linked by disulfide bonds, *J. Exp. Med.* 153: 1503 (1981).

20. P. Gunning, J. Leavitt, G. Muscat, S. Ng, and L. Kedes, A human β-actin expression vector system directs high-level accumulation of antisense transcripts, *Proc. Natl. Acad. Sci.* 84:4831 (1987).

21. S.D. Gorman, B. Tourvieille, and J.R. Parnes, Structure of the mouse gene encoding CD4 and an unusual transcript in brain, *Proc. Natl. Acad. Sci. U.S.A.* 84:7644 (1987).

22. R.H. Seong, J.W. Chamberlain, and J.R. Parnes, CD4 transmembrane region and/or cytoplasmic tail delivers a specific signal for T cell differentiation to a CD4 cell lineage, *Nature* In press.

23. P. Kisielow, H. Bluthman, U.D. Staerz, M.Steinmetz, and H. von Boehmer, Tolerence in T-cell-receptor transgenic mice involves deletion of nonmature CD4$^+$CD8$^+$ thymocytes, *Nature* 333:742 (1988).

24. P. Kisielow, S.-H. Teh, H. Bluthman, and H. von Boehmer, Positive selection of antigen-specific T cells in thymus by restricting MHC molecules, *Nature* 335:730 (1988).

25. M.J. Bevan and P. Fink, The influence of thymus H-2 antigens on the specificity of maturing killer and helper cells, *Immunol. Rev.* 269:417 (1977).

26. R.H. Zinkernagel, G. Callahan, A. Althage, S. Cooper, P. Klein, and J. Klein, On the thymus in the differentiation of H-2 self-recognition by T cells: evidence for dual recognition? *J. Exp. Med.* 147:882 (1978).

27. H. von Boehmer, W. Haas, and N.K. Jerne, Major histocompatibilitycomplex-linked immune-responsiveness is aquired by lymphocytes of low-responder mice differentiating in thymus of high-responder mice, *Proc. Natl. Acad. Sci. U.S.A.* 75: 2434 (1978).

28. J. Kappler and P.J. Marrack, The role of H-2 linked genes in helper T cell function IV. Importance of T-cell genotype and host environment in I region and Ir gene expression, *J. Exp. Med.* 148:1510 (1978).

29. C.E. Rudd, J.M. Trevillyan, J.D. Dasgupta, L.L. Wong, and S.F. Schlossman, The CD4 receptor is complexed in detergent lysates to a protein -tyrosine kinase (pp58) from human T lymphocytes, *Proc. Natl. Acad. Sci. U.S.A.* 85:5190 (1988).

30. A. Veillete, M.A. Bookman, E.M. Horak, and J.B. Bolen, The CD4 and CD8 T cell surface antigens are physically associated with the internal membrane tyrosine kinase p56[lck], *Cell* 55:301 (1988).

31. A. Veillete, J.C. Zuniga-Pflucker, J.B. Bolen, and A.M. Kruisbeek, Engagement of CD4 and CD8 expressed on immature thymocytes induces activation of intracellular tyrosine phosphorylation pathways, *J. Exp. Med.* 170:1671 (1989).

DIFFERENTIAL INVOLVEMENT OF PROTEIN TYROSINE KINASES
p56[lck] and p59[fyn] IN T CELL DEVELOPMENT

Nicolai S.C. van Oers, Alex M. Garvin [Δ], Michael P. Cooke *, Craig B. Davis [†],
D.R. Littman[†], Roger M. Perlmutter *, and Hung-Sia Teh

Department of Microbiology, University of British Columbia, Vancouver, Canada V6T
1Z3 ; the [Δ]CIML, case 906, 13288 Marseille, Cedex 9, France; the *Departments of
Immunology, Biochemistry, and Medicine (Medical Genetics), and the Howard
Hughes Medical Institute, University of Washington, Seattle, WA 98195, and the
[†]Howard Hughes Medical Institute, Department of Microbiology and Immunology,
University of California, San Francisco, San Francisco, CA 94143.

INTRODUCTION

Mature CD4+ and CD8+ T cells recognize and respond to processed antigens associated
with major histocompatibility complex-encoded molecules (MHC). It is during thymic
development that selection processes act on immature CD4+CD8+ thymocytes to ensure the
formation of a repertoire of functional T cells[1,2]. Thus, immature T cells expressing an αβ
TCR with specificity for self-peptides plus MHC class I or class II molecules are positively
selected, differentiating into CD8+ or CD4+ T cells, respectively[3,4,5,6]. Immature
thymocytes lacking or expressing a non-selectable TCR undergo programmed cell death[1].
Additionally, those CD4+CD8+ T cells expressing an autospecific TCR may undergo
programmed cell death[7,8] (negative selection). The CD4 and CD8 molecules are thought to
actively participate in these T cell repertoire selection events through their coreceptor
functions[9]. In a coreceptor model, the TCR and appropriate coreceptor molecule recognize
and interact with the same MHC molecule, with CD4 or CD8 potentiating TCR-derived
intracellular signals[10,11,12]. Recent reports have demonstrated that mutations affecting
CD8/class I MHC interactions disrupt both positive and negative selection in the
thymus[13,14].

TCR tiggering in mature T cells activates multiple biochemical pathways including protein tyrosine kinase(s) and phosphoinositide-specific phospholipase C, leading to the generation of numerous second messengers[15]. Since the TCR/CD3 complex possesses no known enzymic functions, signal transduction following TCR/coreceptor/ligand interactions is proposed to result at least in part from coupled protein tyrosine kinases (PTK). One recently identified PTK potentially associated with the TCR is a unique alternatively spliced form of the proto-oncogene, p59^{fyn} [16,17]. Interestingly, the expression of p59^{fyn} in thymocytes is developmentally regulated since CD4$^+$CD8$^+$ T cells express levels of p59^{fyn} approximalely 10-fold lower than mature CD4$^+$ and CD8$^+$ T cells[18]. Experimental augmentation of p59^{fyn} expression in immature thymocytes results in enhanced intracellular signalling following TCR ligation, suggesting that this PTK is involved in thymocyte development[18].

In addition to the TCR associated PTK(s), both the CD4 and CD8 coreceptor molecules physically associate with another PTK, p56^{lck} [19,20]. Signals generated through the coreceptor molecules are proposed to enhanced TCR-mediated signal transduction[12,21]. In immature thymocytes, cross-linking of surface CD4 or CD8 alone or in conjunction with the TCR potentiates the kinase activity of p56^{lck}, resulting in the activation of cellular protein tyrosine kinase pathways[21,22]. Furthermore, high levels of p56^{lck} expression in transgenic mice severely disrupt thymopoeisis[23]. Taken together, these data suggest that the TCR and the coreceptor associated PTKs are critical components in regulating thymocyte development. However, the contribution of the TCR and coreceptor associated PTKs and their respective signalling functions in T cell repertoire selection processes remains to be established.

To address the contribution of the TCR and coreceptor-derived signals during T cell repertoire selection, we have generated transgenic mice which have augmented expression of either CD4 or p59^{fyn} in immature thymocytes. These mice were mated with transgenic lines expressing an $\alpha\beta$ TCR specific for the male (H-Y) antigen presented by H-2D^b class I MHC molecules. As summarized here, the augmented expression of CD4 molecules in immature thymocytes disrupts both CD8-dependent positive and negative selection[24,25]. Interference of positive and negative selection was dependent on the presence of an intact CD4 molecule capable of interacting with p56^{lck}. This finding was correlated with the observation that increased levels of CD4 reduce the percentage of total p56^{lck} associated with CD8. Furthermore, we find that approximately 5-10-fold increased levels of p59^{fyn} (TF) expression in immature thymocytes fail to modify the outcome of CD8-dependent selection events in $\alpha\beta$/TF double transgenic mice. Therefore, although p59^{fyn} overexpression contributes to enhanced TCR signalling in immature thymocytes, enhanced levels of this PTK have no obvious effects on T cell development. In contrast, thymocyte development appears to be tightly regulated by the coreceptor associated PTK p56^{lck}.

MATERIALS AND METHODS

Mice

C57L and C57BL/6 mice were purchased from Jackson Laboratories (Bar Harbor, Maine) and bred in the animal facility at the University of British Columbia. H-2^b $\alpha\beta$ TCR transgenic mice, *lck*-CD4 transgenic mice (full-length CD4 [line 727] and CD4 tailless [line 3621]), and *lck*-p59*fyn* transgenic mice (TF, thymic form of p59*fyn*) were produced as previously described and bred in our facility [8,18,24,25].

Antibody Preparations and Flow Cytometry

Monoclonal antibodies (mAb) directed against the α (T3.70) and β (F23.1) TCR transgene product were prepared as previously described[3,26] while fluorescein-conjugated anti-CD8 and phycoerythrin-conjugated anti-CD4 were purchased from Becton-Dickinson (Mississauga, Ontario). The anti-*lck* antiserum used in these studies is a rabbit antiserum generated against a TrpE-*lck* fusion protein and was kindly provided by Dr. A. Veillette (anti-*lck* 7229, McGill University, Montreal). Thymocytes from mice ranging in age from 1-8 weeks were prepared as previously described and labeled with fluorochrome conjugated mAb directed against CD4, CD8 or the α and β TCR transgene products[24,25]. Five to fifteen thousand labeled cells were analyzed with a FACScan flow cytometer (Becton-Dickinson) and the data computed with FACScan research software programs.

Immunoprecipitations and Immunoblotting

Thymocytes from the different mice were isolated and lysed at 5×10^7 cells/ml in TNE (20mM Tris-Cl, pH 8.0; 150 mM NaCl; 2mM EDTA) with either 1% NP-40 or 1% Triton-X-100 with protease and phosphatase inhibitors at 4°C for 15 min. The percentage of total cellular p56*lck* coimmunoprecipitating with CD4 or CD8 was determined as previously described[25]. For immunoblotting, total cell lysates from equal cell numbers (3×10^6) were resolved on 8% SDS-PAGE and Western blotted as previously described[25]. Polyclonal rabbit anti-*fyn* antibodies and monoclonal anti-phosphotyrosine antibodies (4G10) were purchased from Upstate Biotechnology (UBI, Lake Placid, New York) and used according to the suppliers instructions.

RESULTS

Augmented Expression of Full-length CD4 Molecules in Immature Thymocytes Impairs CD8-dependent Positive and Negative Selection

To characterize the effects of CD4 overexpression on T cell repertoire selection, transgenic mice expressing full-length CD4 molecules (727) and CD4 molecules with a truncated cytoplasmic domain (3621) were mated with transgenic mice expressing an αβ TCR specific for the male H-Y peptide presented by H-2D[b] class I MHC molecules. CD4 molecules with a deleted cytoplasmic domain are unable to associate with the PTK, p56*lck*[27,28]. As previously reported, the surface expression of CD4 on thymocytes from the 2 transgenic lines 727 and 3621 is elevated 10-15 fold (Figure 1)[24,25]. Since both CD4 and CD8 coreceptor molecules associate with p56*lck*, it is expected that increased expression of intact, but not truncated CD4, may lead to a decreased association of the CD8 coreceptor with p56*lck*.

The majority of autospecific immature CD4+CD8+ thymocytes in αβ TCR transgenic

91

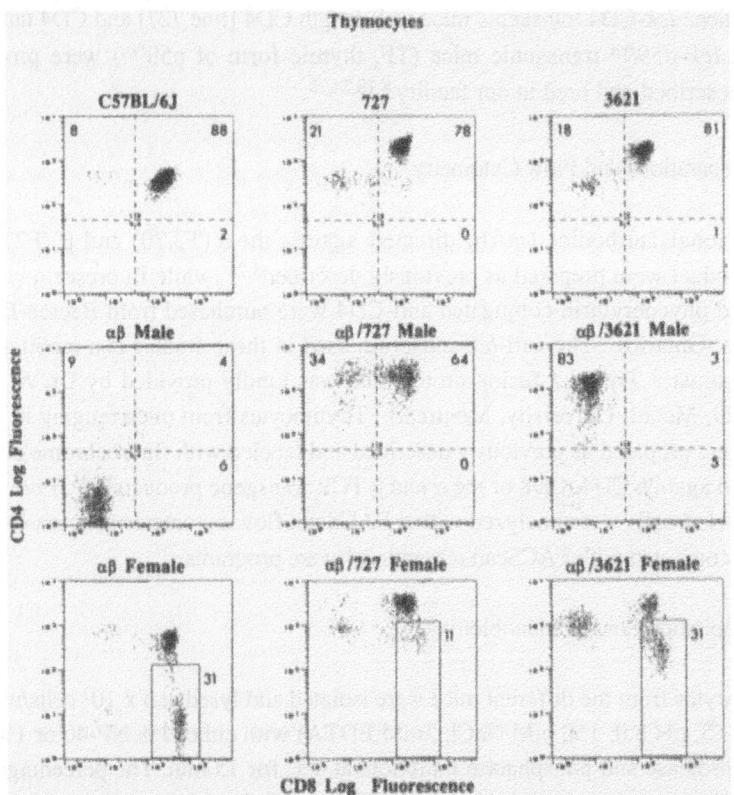

Figure 1. Representative 2-color analyses of CD4 and CD8 surface expression in αβ transgenic, CD4 transgenic (727 and 3621), and αβ/CD4 double transgenic mice (αβ/727 and αβ/3621) male and female mice. Thymocytes from 4-8 week old C57BL/6J , 727, 3621, αβ/727 and αβ/3621 mice were collected and stained with mAb specific for CD4 and CD8. The numbers in the quadrants represent the percentage of cells in each population. For female transgenic mice, the percentage of CD8+ thymocytes are shown in the boxed in region. A total of 10,000 events/sample were analyzed with a FACScan Flow cytometer and the data computed using FACScan software programs (Becton-Dickinson). The data presented in this Figure is modified from the results published in van Oers et al[25].

male mice are deleted by a process of accelerated programmed cell death in a mechanism involving interactions between the αβ TCR, MHC Class I H-2D^b molecules and CD8, as well as LFA-1/ICAM interactions[4,29]. As a consequence of this clonal deletion, the residual population of T cells in the thymus of these mice exhibit a CD4-CD8- αβ TCR+ phenotype. As shown in Figure 1, the introduction of augmented levels of full-length CD4 in the immature thymocytes of these mice (αβ/727) leads to a significant increase in percentage of CD4+CD8+ T cells. In addition, there is an increase in the total number of thymocytes with the majority of CD4+CD8+ T cells maintaining expression of the autospecific αβ TCR (data not shown). Protection from clonal deletion requires the overexpression of CD4 molecules capable of signal transduction since the elevated expression of cytoplasmically truncated CD4 molecules does not affect CD8-dependent negative selection (αβ/3621 male mice).

Table 1. Percentage of p56^{lck} co-immunoprecipitating with the CD4 and CD8 coreceptor molecules.

Thymocytes	Percentage of total p56^{lck}	
	CD4	CD8
C57BL/6J	30%	17%
727	55%	3%
3621	27%	13%

[1] Detergent lysates of thymocytes isolated from the indicated mice were prepared and immunoprecipitated with anti-*lck*, anti-CD4 (GK1.5) or anti-CD8 (53.67). The percentage of p56^{lck} co-immunoprecipitating with CD4 or CD8 was calculated as described in Materials and Methods.

We then examined the effects of wild-type and truncated CD4 surface expression on CD8-dependent positive selection. As previously described[3], the specificity of the αβ TCR for the thymic MHC class I H-2D^b molecule in female mice results in the preferential production of CD8$^+$ T cells expressing the αβ transgenic TCR. Shown in Figure 1, and as reported elsewhere[24,25], the positive selection of mature CD8$^+$ T cells is inhibited in double transgenic female mice (αβ/727) expressing augmented levels of CD4 molecules in immature thymocytes. This is illustrated by the reduced percentage of CD8$^+$ T cells as delineated by the boxed-in area. Again, impaired positive selection is contingent on full-length CD4 expression since the presence of truncated CD4 molecules (αβ/3621) fails to influence the development of CD8$^+$ T cells. In addition, only approximately 30% of the CD8$^+$ T cells that do develop in αβ/727 mice maintain transgenic TCR α expression compared to 95% for αβ and αβ/3621 CD8$^+$ T cells[24,25].

Augmented CD4 Expression Influences Coreceptor/p56^{lck} Interactions

The cytoplasmic domains of CD4 and CD8α are associated with the PTK, p56^{lck} [27,28]. The aforementioned results suggested that the augmented expression of full-length but not tailless CD4 molecules may have influenced TCR/CD8 coreceptor signalling events. To examine this possibility, we assessed the percentage of p56^{lck} associated with the coreceptor molecules in C57BL/6J mice and the transgenic lines 727 and 3621. As summarized in Table I, overexpression of CD4 in immature thymocytes contributes to a 1.5-2 fold increase in the percentage of p56^{lck} associated with CD4. Furthermore, a concomitant 2-3 fold decrease in the percentage of p56^{lck} coimmunoprecipitating with CD8 was observed. These data would suggest that interference with positive and negative selection by CD4 overexpression may be in part attributable to decreased CD8/p56^{lck}-derived signals and increased CD4/p56^{lck}-mediated intracellular signals.

Consequences of p59^{fyn} Overexpression on Thymopoiesis

To examine the effect of p59^{fyn} overexpression on CD8-dependent positive and negative selection, we generated transgenic mice expressing murine p59^{fyn} under the control of the *lck* proximal promoter. The transgenic mice expressing the thymic form of p59^{fyn} (TF) were mated with the αβ TCR trangenic mice. Western blot analyses of thymus cell lysates from the αβ/TF transgenic mice revealed a 5-10-fold increase in p59^{fyn} expression and increased protein tyrosine phosphorylation patterns[18] (Figure 2). As previously reported, the phenotypic changes associated with deletion in αβ TCR transgenic male mice are apparent in embryonic thymuses from day 18 onwards[30]. To assess whether elevated levels of p59^{fyn} in immature thymocytes led to more efficient delivery of deleting signals, we examined αβ/TF double transgenic mice at 1 week and 8 weeks of age. Thymocytes from 1 week old and adult double transgenic male mice (αβ/TF) were analyzed for the surface expression of CD4, CD8 and the β and α TCR trangene products. As shown in Figure 3, phenotypic changes associated with deletion in αβ TCR transgenic male mice are clearly evident by 1 week of age, as denoted by the reduction of CD4$^+$CD8$^+$ T cells. Augmented expression of p59^{fyn} in these mice (αβ/TF) fails to accelerate the normal developmental sequence of clonal deletion. Furthermore, the level of expression of the β and α transgene products remains very similar to that noted for αβ TCR transgenic male mice. By 8 weeks of age, the majority of CD4$^+$CD8$^+$ cells in the αβ male mice are eliminated, and these patterns of deletion are unaffected by the elevated expression of p59^{fyn}.

Since the phenotypic changes associated with positive selection are not evident until 4-6 weeks of age[30], it was also of interest to determine whether augmented p59^{fyn} expression influenced the normal ontogeny of positive selection. As illustrated in Figure 4, by 1 week of age approximately 6% of thymocytes from αβ female mice are of the CD4$^-$CD8$^+$ phenotype. By 8 weeks, between 15-40% of total thymocytes can be of the CD8$^+$ phenotype with the vast majority maintaining expression of the αβ TCR[1,3]. To ascertain whether p59^{fyn} overexpression could alter the kinetics or outcome of positive selection, we analyzed αβ/TF double trangenic female mice. The findings in Figure 3 indicate that the augmented

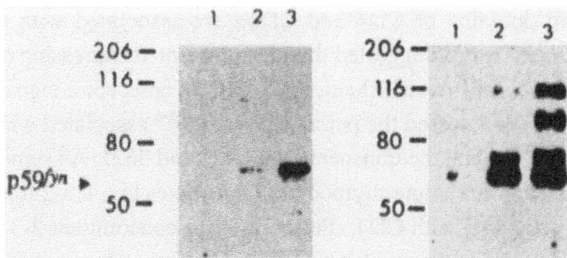

Figure 2. p59^{fyn} expression and phosphotyrosine containing proteins in C57BL/6J, αβ , and αβ/TF thymocytes from male mice. Whole cell lysates from C57BL/6J, αβ and αβ/TF thymocytes were immunoblotted with antibodies against p59^{fyn} (Panel A) or an anti-phosphotyrosine mAb (Panel B) as described in Materials and Methods. Lane 1, C57BL/6J thymocytes; Lane 2, thymocytes from αβ TCR transgenic male mice; Lane 3, thymocytes from αβ/TF double transgenic male mice.

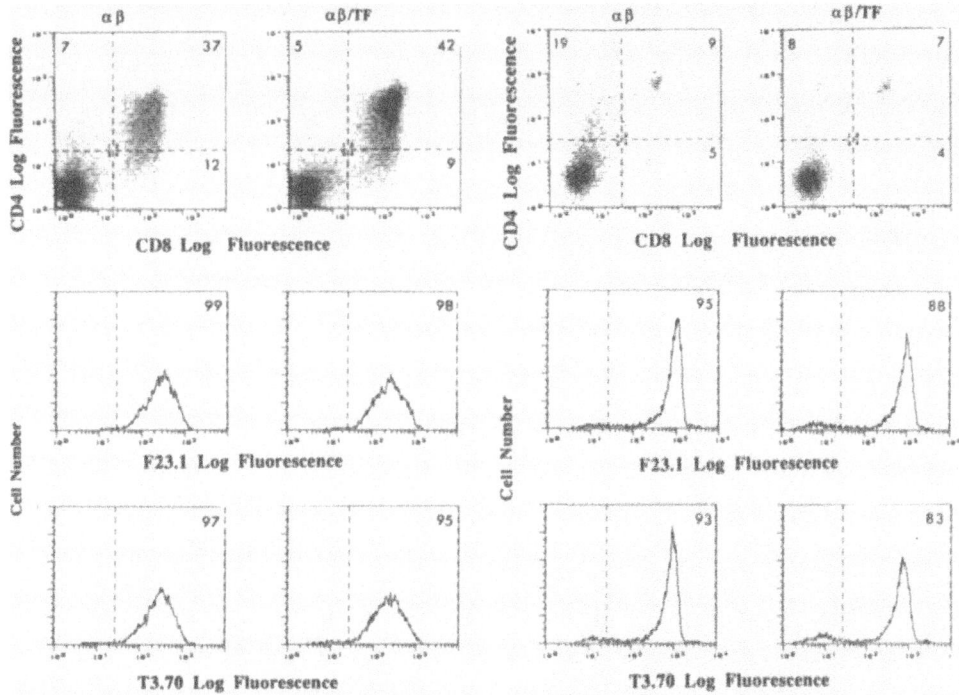

Figure 3. Ontogeny of T cells in single transgenic (αβ) and double transgenic (αβ/TF) male mice. Thymocytes from the indicated age were collected and stained with phycoerythrin-labeled CD4 and fluorescein-labeled CD8 mAb and 10,000 cells/sample were analyzed with a FACScan flow cytometer. For TCR staining, thymocytes were labeled with mAbs specific for the β (F23.1) and α (T3.70) transgene product followed by a Goat-anti-Mouse IgG conjugated to FITC.

p59^{fyn}levels did not contribute to an accelerated appearance of positively selected cells at early developmental time points. Furthermore, the percentage of mature CD8+ T cells in adult female αβ/TF mice is comparable to that found in control αβ TCR transgenic female mice. These findings are consistent with the interpretation that overexpression of a potential TCR associated PTK, p59^{fyn} does not markedly interfere with TCR-mediated positive selection.

DISCUSSION

To investigate the functional contribution of the TCR- and coreceptor-associated PTKs in T cell repertoire selection, transgenic animals with augmented expression of CD4 and p59^{fyn} were mated with αβ TCR transgenic mice. In this report, we provide evidence that the PTK p56^{lck} but not p59^{fyn} appears to be a critical signalling component in T cell development. First, the overexpression of CD4 molecules in transgenic mice expressing a class I restricted αβ TCR interferes with both CD8-dependent negative and positive selection. Second, the disruption of T cell repertoire selection requires the expression of full-length

95

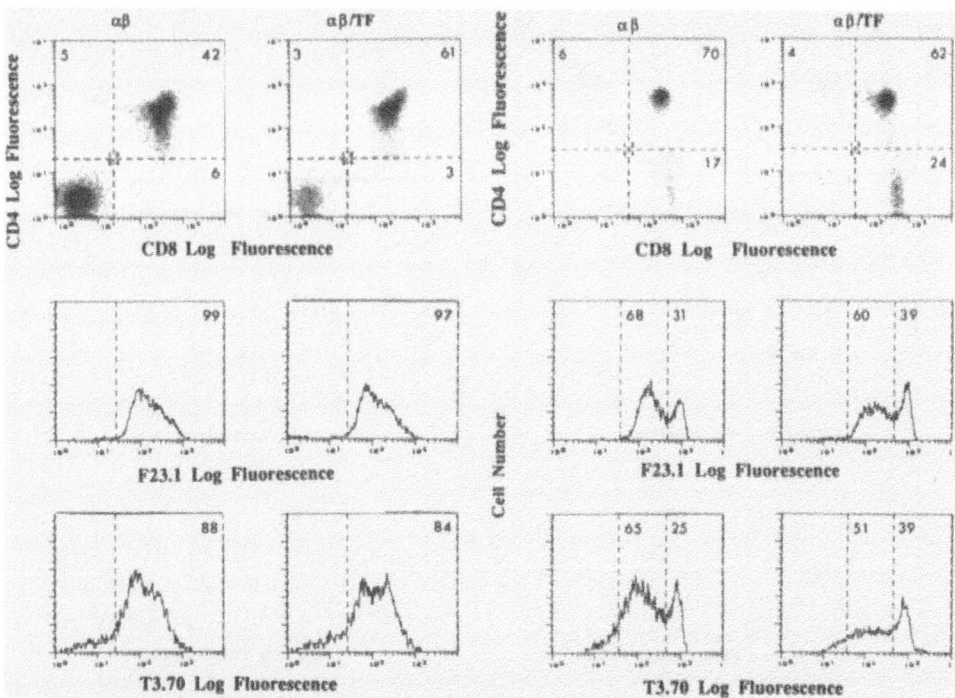

Figure 4. Ontogeny of positive selection in αβ and αβ/TF transgenic female mice. Thymocytes from 1 week and 8 week old mice were collected and stained as described in the legend to Figure 3.

CD4 molecules capable of intracellular signal transduction. Thus, only elevated expression of CD4 molecules capable of associating with the PTK p56^{lck} are able to interfere with normal CD8-dependent selection events. Third, CD4 overexpression leads to an increased percentage of p56^{lck} associating with CD4 while CD8/p56^{lck} interactions are decreased threefold. Taken together, these results would suggest that augmented CD4 levels may in effect attenuate normal CD8/p56^{lck}-derived signals. Alternatively, it is conceivable that CD4 overexpression, contributing to increased CD4/p56^{lck} interactions, may in effect alter normal p56^{lck} regulated developmental processes. For example, p56^{lck} overexpression in transgenic mice severely disrupts normal thymopoeisis[23].

As previously reported, overexpression of p59^{fyn} in thymocytes leads to T cell hyper-responsiveness as assessed by protein tyrosine phosphorylation changes and [Ca^{2+}]$_i$ mobilization following TCR ligation[18]. These effects are unique to p59^{fyn} since elevated expression of p56^{lck} or p59^{hck} in immature thymocytes yields very distinct phenotypic outcomes[23]. Surprisingly, we have found that overexpression of p59^{fyn} does not influence intrathymic selection of H-Y specific T cells. Thus, a 5-10-fold elevated level of p59^{fyn} fails to influence either the kinetics or pattern of CD8-dependent negative and positive selection. There are several possible explanations for the lack of effect of p59^{fyn} overexpression. First, in spite of enhanced intracellular signalling, immature thymocytes overexpressing p59^{fyn} do not acquire mature T cell responsiveness[18]. Indeed, the possibility exists that additional cellular components, necessary for conferring mature T cell responsiveness, are limiting in immature thymocytes. Second, positive and negative selection of cells in the H-Y αβ TCR transgenic mice in the context of H-2^b may already occur at maximum efficiency, such that no augmentation of signalling could alter the selection processes. Finally, other PTKs and/or signalling components may be involved in transducing TCR-mediated signals. For example, a novel PTK has recently been shown to associate with the TCR/CD3-ζ chain[31]. In fact, different components of the TCR/CD3 complex may function as distinct signalling modules, associating with specific PTKs, each potentially exhibiting some selective substrate specificity[32,33]. Overexpression of any one TCR-associated PTK may not be sufficient to perturb normal thymopoiesis. In contrast, the coreceptor associated PTK, p56^{lck} appears to be a critical component operating during T cell repertoire selection.

ACKNOWLEDGEMENTS

The authors would like to thank Dr. Andre Veillette for providing the anti-*lck* antibodies used in these studies. This work was supported by a grant from the Medical Research Council of Canada to H-S.T. and grants from the National Institutes of Health and the Howard Hughes Medical Institue to R.M.P. and D.R.L. N.v.O. was supported by a post-doctoral fellowship from the Natural Sciences and Engineering Research Council of Canada, A.M.G. and M.P.C. by pre-doctoral training grants from NIGMS, and C.B.D. by a post-doctoral fellowship from the Juvenile Diabetes Foundation.

REFERENCES

1. H. von Boehmer, Developmental biology of T cells in T cell receptor transgenic mice, Ann. Rev. Immunol. 8:531 (1990).

2. M. Blackman, J. Kappler, and P. Marrack, The role of the T cell receptor in positive and negative selection of developing T cells, Science 248:1335 (1990).

3. H-S. Teh, P. Kisielow, B. Scott, H. Kishi, Y. Uematsu, H. Bluthmann, and H. von Boehmer, Thymic major histocompatibility complex antigens and the αβ T-cell receptor determine the CD4/CD8 phenotype of T cells, Nature 335:229 (1988).

4. H. von Boehmer, and P. Kisielow, Self-nonself discrimination by T cells, Science 248:1369 (1990).

5. L.J. Berg, A.M. Pullen, B.F. de St. Groth, D. Mathis, C. Benoist, and M. Davis, Antigen/MHC specific T-cells are preferentially exported from the thymus in the presence of their MHC ligand, Cell 58:1035 (1989).

6. W.C. Sha, C.A. Nelson, R.D. Newberry, J.K. Pullen, L.R. Pease, J.H. Russell, and D.Y. Loh, Positive selection of transgenic-receptor bearing thymocytes by K^b antigen is altered by K^b mutations that involve peptide binding, Proc. Natl. Acad. Sci. 87:6186 (1990).

7. J.W. Kappler, N. Roehm, and P. Marrack, T cell tolerance by clonal elimination in the thymus, Cell 49:273 (1987).

8. P. Kisielow, H. Bluthman, U.D. Staerz, M. Steinmetz, and H. von Boehmer, Tolerance in T cell receptor transgenic mice involves deletion of non-mature CD4$^+$CD8$^+$ thymocytes, Nature 333:742 (1988).

9.   C.A. Janeway, Jr., Accessories of coreceptors, Nature 335:208 (1988).

10. U. Diazani, A. Shaw, B.K. Al-Ramadi, R.T. Kubo, and C.A. Janeway, Jr., Physical association of CD4 with the T cell receptor, J. Immunol. 148:678 (1992).

11. R.D. Salter, R.J. Benjamin, P.K. Wesley, S.E. Buxton, T.P.J. Garrett, C. Clayberger, A.M. Krensky, A.M Norment, D.R. Littman, and P. Parham, A binding site for the T -cell coreceptor CD8 on the α_3 domain of HLA-A2, Nature 345:41 (1990).

12. F. Emmrich, V. Strittmatter, and K. Eichmann, Cross-linking of the TCR complex with the subset-specific differentiation antigen stimulates interleukin 2 receptor expression in human CD4 and CD8 T cells, Proc. Natl. Acad. Sci. 83:8298 (1986).

13. A.L. Ingold, C. Landel, C. Knall, G.A. Evans, and T. Potter, Co-engagement of CD8 with the T cell receptor is required for negative selection, Nature 352:721 (1991).

14. C.J. Aldrich, R.E. Hammer, S. Jones-Youngblood, U. Koszinowski, L. Hood, I. Stoyonowski, and J. Forman, Negative and positive selection of antigen-specific cytotoxic T lymphocytes affected by the $\alpha3$ domain of MHC classs I molecules. Nature 352:718 (1991).

15. A. Weiss, J. Imboden, K. Hardy, B. Manger, C. Terhorst, and J. Stobo, The role of the T3/antigen receptor complex in T cell activation, Ann. Rev. Immunol. 4:593 (1986).

16. M.P. Cooke, and R.M. Perlmutter, Expression of a novel form of the fyn proto-oncogene in hematopoeitic cells, New Biologist 1:66 (1989).

17. L.E. Samelson, A.F. Phillips, E.T. Luong, and R.D. Klausner, Association of the fyn protein tyrosine kinase with the T cell antigen receptor, Proc. Natl. Acad. Sci. 87:4358 (1990).

18. M.P. Cooke, K. M. Abraham, K.A. Forbush, and R.M. Perlmutter, Regulation of T cell receptor signalling by a *src* family protein tyrosine kinase (p59^{fyn}), Cell 65:281 (1991).

19. C. E. Rudd, J. M. Trevillyan, J.D. Dasgupta, L.W. Wong, and S.F. Schlossman, The CD4 receptor is complexed in detergent lysates to a protein tyrosine kinase (pp58) from human T lymphocytes, Proc. Natl. Acad. Sci. 85:5190 (1988).

20. A. Veillette, M.A. Bookman, E.M. Horak, and J.B. Bolen, The CD4 and CD8 T cell surface antigens are associated with the internal membrane tyrosine-protein kinase p56^{lck}, Cell 55:301 (1988).

21. A. Veillette, J.C. Zuniga-Pflucker, J.B. Bolen, and A.M. Kruisbeek, Engagment of CD4 and CD8 expressed on immature thymocytes induces activation of intracellular tyrosine phosphorylation pathways, J. Exp. Med. 170:1671 (1989).

22. L.K. Gilliland, H-S. Teh, F.M. Uckun, N.A. Norris, S-J. Teh, G.L. Schieven, and J.A. Ledbetter, CD4 and CD8 are positive regulators of T cell receptor signal transduction in early T cell differentiation, J. Immunol. 146:1759 (1991).

23. K.M. Abraham, S.D. Levin, J.D. Marth, K.A. Forbush, and R.M. Perlmutter, Delayed thymocyte development induced by augmented expression of p56^{lck}, J. Exp. Med. 173: 1421 (1991).

24. H-S. Teh, A.M. Garvin, K.A. Forbush, D.A. Carlow, C.B. Davis, D.R. Littman, and R.M. Perlmutter, Participation of CD4 coreceptor molecules in T cell repertoire selection, Nature 349: 241 (1991).

25. N.S.C. van Oers, A.M. Garvin, C.B. Davis, K.A. Forbush, D.A. Carlow, D.R. Littman, R.M. Perlmutter, and H-S. Teh, Disruption of CD8-dependent negative and positive selection of thymocytes is correlated with a decreased association between CD8 and the protein tyrosine kinase, p56^{lck}, Eur. J. Immunol. In Press(1992).

26. U. Staerz, H. Rammensee, J. Benedetto, and M. Bevan, Characterization of a murine monoclonal antibody specific for an allotypic determinant on T cell antigen receptor, J. Immunol. 134:3994 (1985).

27. A.S. Shaw, K.E. Amrein, C. Hammond, D.F. Stern, B.M. Sefton, and J.K. Rose, The lck tyrosine

protein kinase interacts with the cytoplasmic tail of the CD4 glycoprotein through its unique amino-terminal domain, Cell 59:627 (1989).

28. J.M. Turner, M.H. Brodsky, B.A. Irving, S.D. Levin, R.M. Perlmutter, and D.R. Littman, Interaction of the unique N-terminal region of tyrosine kinase p56[lck] with cytoplasmic domains of CD4 and CD8 is mediated by cysteine motifs, Cell 60:755 (1990).

29. D.A. Carlow, N.S.C van Oers, S-J. Teh, and H-S. Teh, Deletion of antigen-specific immature thymocytes by dendritic cells requires LFA-1/ICAM interaction, J. Immunol. In Press (1992).

30. H-S. Teh, H. Kishi, B. Scott, P. Borgulya, H. von Boehmer, and P. Kisielow, Early deletion and late positive selection of T cells expressing a male-specific receptor in T-cell receptor transgenic mice, Develop. Immunol. 1:1 (1990).

31. A. Chan, B.A. Irving, J.D. Fraser, and A. Weiss, The ζ chain is associated with a tyrosine kinase and upon T-cell antigen receptor stimulation associates with ZAP-70, a 70-kDa tyrosine phosphoprotein, Proc. Natl. Acad. Sci. 88:9166 (1991).

32. F. Letourneur, and R.D. Klausner, Activation of T cells by a tyrosine kinase activation domain in the cytoplasmic tail of CD3-ε, Science 255:79 (1992).

33. A-M. K. Wegener, F. Letourner, A. Hoeveler, T. Brocker, F. Luton, and B. Malissen, The T cell receptor/CD3 complex is composed of at least two autonomous transduction modules, Cell 68:83 (1992).

tyrosine kinase interacts with the cytoplasmic tail of the CD4 glycoprotein through its amino-terminal domain. Cell 59:627 (1989).

MECHANISM OF TOLERANCE INDUCTION

Ada M. Kruisbeek, John D. Nieland, and Lori A. Jones

Division of Immunology
The Netherlands Cancer Institute
Plesmanlaan 121
1066 CX Amsterdam, The Netherlands

GENERAL INTRODUCTION

The randomly generated T-cell repertoire is shaped by two processes that occur in the thymus, positive and negative selection (1-5). Positive selection is responsible for generating a T cell repertoire that has the ability to recognize antigenic peptides in association with self-major histocompatibility complex (MHC) molecules. The other process is called negative selection, and assures that tolerance for self-antigens is achieved. During negative selection, potentially autoreactive T cells (i.e. those with high affinity for "self" antigens presented on self-MHC molecules) are actually deleted from the T cell repertoire (2,6,7) or are clonally inactivated (1,8). "Self" is defined here as those self or foreign antigens which are present in the thymus at the moment of selection, a prime moment for which appears to be the early neonatal period: Neonatally thymectomized mice develop a variety of tissue-specific autoimmune diseases later in life (9,10). Although factors controlling these autoimmune diseases are poorly understood, defects in clonal inactivation and clonal deletion have been assigned to be the major cause of autoimmune diseases (9,11,12). In the context of the present definition of self, neonatal virus infection of the thymus results in specific viral peptides being presented as self-antigens, resulting in tolerance for foreign peptides (see below).

Mechanisms of Lymphocyte Activation and Immune Regulation IV: Cellular Communications
Edited by S. Gupta and T.A. Waldmann, Plenum Press, New York, 1992

101

Clonal deletion is the predominant pathway to achieve this neonatal tolerance: T cells expressing particular Vβ chains of the αβ-T cell receptor (TCR) are eliminated or greatly diminished in the periphery of mice expressing a specific combination of self-MHC and self antigens defined as "superantigens". Endogenous antigens thus far known as Mls (minor lymphocyte stimulatory) belong to the class of superantigens, and are now known to be encoded by an open reading frame (orf) in the 3' LTR of mouse mammary tumor virus (MMTV) (13-16). Several correlations between expression of superantigens on negative selecting elements and Vβ chain deletions have been identified. These correlations include: Mls 1[a]: Vβ 6, 7, 8.1, 9 (6,7,17,18); Mls 2[a]: Vβ 3 (19); Mls 3[a]: Vβ 3, 5 (13,14); Mls 4[a]: Vβ 3 (13); Etc-1: Vβ 5.1, 5.2, 11 (15,16,20); Dvb 11.1: Vβ 11 (16); Dvb11.3: Vβ 11 (16); EαEβ plus as yet undefined self peptide(s): Vβ 12, 17 (2,20-22). Clonal deletion also appears to be the predominant mechanism of tolerance acquisition for conventional antigens such as HY and ovalbumin, as demonstrated in TCR-transgenic mice models of von Boehmer et al. (23) and Murphy et al. (24). The removal of these potentially autoreactive T cells from the T cell repertoire appears to happen either at the late double positive (DP) stage in T cell development (i.e., on cells expressing both CD4 and CD8) or at the early single positive (SP) stage (i.e., on cells expressing only CD4 or CD8) (2). Several labs including our own identified other mechanisms involved in maintaining T cell tolerance, i.e., induction of clonal unresponsiveness (anergy) (1,3,8). Thus, T cells that escaped intrathymic clonal deletion can be rendered nonfunctional in another way.

These studies on the mechanisms of tolerance acquisition suggest a variety of approaches for manipulation of the immune system: how can an unwanted immune response against self-antigens or foreign antigens be prevented, and how can unwanted tolerance for pathogens be broken? The present review summarizes how induction or breakdown of immunological tolerance may be achieved in a variety of experimental models of tolerance.

INDUCTION OF PERIFERAL IMMUNOLOGICAL TOLERANCE

T cells with self-reactive antigen receptors are usually removed through intrathymic clonal deletion (1,2). If, however, that process fails for any number of reasons (i.e., a self-antigen is not expressed in the thymus, or expressed on non-hemopoietic antigen presenting cells, APC), other safeguards are built in: we recently demonstrated *in vivo* clonal non-responsiveness in T cells (anergy) with self-reactive receptors in a variety of models in which clonal deletion was precluded (1,2). Current concepts on immunological tolerance hold that anergy is the result of absence of a so-called "co-stimulatory signal" in T cells in which TCR/MHC-peptide interactions occur (25-27). This rather ill-defined co-stimulatory signal is derived from APC. T cells stimulated with antigenic peptides and metabolically inactive or non-bone marrow derived APC become anergic, rather than activated. Thus, one of the ways in which *in vivo* tolerance for self-peptides may be maintained is through recognition of self-peptides on APC incapable of providing co-stimulatory signals. The best candidate co-stimulatory signal that determines whether TCR-occupancy leads to a productive immune response is that generated by the interaction of CD28 (on T cells) and its ligand B7 (on APC) (28,29): anti-CD28 crosslinking is the only intervention identified today that can replace APC, and bypass anergy induction (26,30). Furthermore, mAb against human B7 blocks generation of co-stimulatory signals in human B cells (31). Obviously, blocking of CD28-B7 interactions might present a powerful immunosuppressive intervention, with the potential to control undesirable immune responses against self-antigens or transplantation antigens.

Another way to induce tolerance in adult mice is to treat mice with superantigens. On stimulation with superantigens, T cells with restricted Vβ's will become clonally deleted or clonally inactivated. Induction of T cell tolerance by superantigens may therefore have therapeutic applications: Kim et al. (32) injected MRL/lpr autoimmune mice were with Staphylococcus enterotoxin B, and this intervention delayed the onset of autoimmune pathology via deletion of Vβ 8$^+$ double negative T cells.

INDUCTION OF NEONATAL TOLERANCE WITH SYNTHETIC PEPTIDES

As discussed above, tolerance for self-antigens is achieved in part during early T cell development, and may occur either by clonal deletion or by induction of clonal anergy. Autoimmune diseases are often a consequence of failures in the selection of the T cell repertoire; clonal deletion may have been ineffective, or the state of anergy for self-antigens in T cells is broken (9-10,33). Yet, negative selection may also have unwarranted side effects: responses to foreign antigens during adult life may be precluded because of their similarity to self-antigens, or because the foreign antigen is already present neonatally, such that the particular antigen is mistakingly regarded as a self-antigen. In agreement with this concept, neonatal exposure to a variety of class II-MHC associated foreign peptides induces immmunological tolerance (34-36). Presumably, the interaction of developing T cells with such mock-self-antigens results in inactivation or deletion. Thus, when at this early stage of development a pathogen would enter the immune system, pathogenic peptides will be seen as self-peptides and reacting T cells will be deleted or inactivated. This may well be the explanation for the observation that neonatal infection with viruses such as Gross Murine Leukemia virus (37) and murine mammary tumor virus (38) renders mice unable to reject tumors arising from exposure to such viruses.

We have begun testing this hypothesis by analyzing tolerance mechanisms operating for class I-associated peptides: anti-viral CTL recognize short peptides derived from viral proteins (e.g., 39); such peptides must also have the ability to bind to MHC-class I molecules (40, 41). Experiments were carried out to establish the conditions required for the neonatal tolerization of mice through the use of peptides representing the major CTL epitopes in the immune response to Sendai virus and adenovirus. Mice can be protected from a lethal

Sendai virus infection by vaccination with the immunodominant viral peptide (SV16 peptide, representing residues 321-336 of Sendai virus nucleoprotein: HGEFAPGNYPALWSYA), through the action of peptide-specific CTL (39). The SV9 peptide FAPGNYPAL binds most efficiently to K and D molecules of the H-2^b haplotype (40,41), and can also function as a CTL epitope (42). Upon treatment of neonatal mice with the SV9 peptide, effective abrogation of the ability to generate primary *in vitro* CTL responses against SV9 can be achieved (Nieland, Jones and Kruisbeek, in preparation). Even upon *in vivo* priming with SV9, such mice were still unable to generate an SV9 specific CTL response (data not shown). These experiments represent the first demonstration that tolerance for a class I-associated foreign peptides, at least as defined by the inability to generate *in vitro* and *in vivo* CTL responses, can be induced through neonatal exposure, in accordance with what was previously shown for class II-associated antigenic peptides.

How unwarranted tolerance for pathogens that have presented as self antigens during early T cell development can be broken can be learned from various models for induction of autoimmune diseases. For instance, introduction of a high amount of the relevant self peptide or protein may result in pathology, as is the case with the autoimmune disease oophoritis induced by the zona pellucida peptide or protein ZP3 (33). A second example is seen in experiments of Lin et al. (43). Mouse cytochrome C was injected into H2^k mice and failed to generated a response; if, however, human cytochrome C was injected into these mice, they did develop a mouse cytochrome C specific immune response. A related approach is to introduce a synthetic peptide with moderate mutations, and thus induce a response which is cross reactive with the original peptide (44).

Also viral infections can cause autoimmune disease. Ohashi et al. (45) and Oldstone et al. (46) showed that viral infection of pancreatic acinar cells resulted in a CD8$^+$ autoreactive T cell response which caused diabetes. Finally *in vivo* introduction of dendritic cells loaded with self peptides may induce an autoimmune response. From experiments of Boog et al. (47) it is known that dendritic cells can induce responses which can not be induced by injecting spleen or lymph node cells. Overall, these models suggest several strategies for breaking (induced) tolerance for non-self-antigens; these include introduction of the pertinent peptide in large quantities, site directed mutations in self peptides, or the use of specific stimulator cells in these strategies.

INDUCTION OF PRIMARY CTL RESPONSES AGAINST SYNTHETIC PEPTIDES

Ability to generate primary CTL responses against synthetic peptides is a prerequisite for the experiments described above: to test whether any particular intervention has indeed generated immunological tolerance, *in vitro* assays that test ability to generate primary CTL responses are required, besides the more conventional *in vitro* assays that utilize T cells from primed mice. These assay systems set the stage for testing the efficacy of the various tolerance protocols described above.

Soluble peptides in the extracellular fluids are excluded from the class I MHC-restricted pathway of antigen presentation (48). However, exogenous peptides can be presented in association with class I MHC molecules on certain specialized APC: DC were recently shown to present both SV9 and A10 peptides in such an effective fashion that even T cells from unprimed mice are induced to display peptide-specific CTL reactivity (49). Presumably, the superior ability of DC to perform as APC for foreign antigenic peptides relies not only on their higher level of class I-MHC expression, but also on their ability to form clusters with T cells. Also the antigen processing-defective cell line RMA-s can induce primary CTL responses in unprimed T cells against SV9 and A10 peptides (42), albeit for quite different reasons than DC: RMA-S cells are superior to the parent RMA cells with respect to peptide-binding ability (49).

FUTURE DIRECTIONS

Developing T cells undergo a process of negative selection to assure that tolerance for self-antigens is achieved, and both clonal deletion and induction of clonal non-responsiveness (so-called "anergy") have been implicated as the major mechanisms by which self-tolerance is acquired. Earlier studies defined conditions in which clonal deletion was precluded or clonal anergy was broken, resulting in autoimmune diseases. One of the unwarranted consequences of this negative selection proces may be that tolerance for foreign antigens results as well: the similarity of foreign antigens to self-antigens, or neonatal presence of foreign antigens, effectively results in those foreign antigens being regarded as self-antigens. Peptides encoded for by tumor viruses appear subject to the same rules: similarity with self-peptides or neonatal infection was shown, for several viruses, to induce specific immunological tolerance in the pertinent antigen-specific T cell population. The hypothesis presented here (i.e., systemic tolerance for certain tumor antigens may preclude an effective immune response) is based on the observation that neonatal infection with tumor viruses will induce tolerance for the pertinent viral antigens, such that tolerized mice will develop tumors later in life when exposed to these tumor-inducing viruses. Examples of viruses for which such rules have been established are Gross murine leukemia virus, mammary tumor viruses and lymphocytic choriomeningitis virus.

We propose to utilize existing models for tolerance induction with synthetic peptides, and study the mechanisms by which specific tolerance can be broken.Goals to be achieved include: 1. to define the conditions for induction of immunological tolerance in neonatal and adult mice, using well-defined synthetic peptides representing the major T cell epitopes of various pathogens; 2. to develop approaches that result in a breakdown of immunological tolerance for foreign antigenic peptides, with the intent of inducing an immune response against pathogens such as viruses. The resulting information may define new interventions that change a condition of immunological non-responsiveness for foreign antigens into one on which an effective immunological response is obtained.

REFERENCES

1. F. Ramsdell, and B.J. Fowlkes, Clonal deletion versus clonal anergy: The role of the thymus in inducing self tolerance, *Science* 248:1343 (1990).
2. J.W. Kappler, N. Roehm, and P. Marrack, T cell tolerance by clonal deletion in the thymus, *Cell* 49:273 (1987).
3. J.C. Zuniga-Pflucker, L.A. Jones, L.T. Chin, and A.M. Kruisbeek, CD4 and CD8 act as co-receptors during thymic selection of the T cell repertoire, *Sem. Immunol.* 3: 167 (1991).
4. P. Kisielow, H.S. Teh, H. Blüthmann, and H.von Boehmer, Positive selection of antigen-specific T cells in thymus by restricting MHC molecules, *Nature* 335:730 (1988).
5. M. Blackman, J. Kappler, and P. Marrack, The role of the T cell receptor in positive and negative selection of developing T cells, *Science* 248:1335 (1990).
6. J.W. Kappler, U.Staerz, J.White, and P.C.Marrack, Self-tolerance eliminates T cells specific for Mls-modified products of the major histocompatibility complex, *Nature* 332:35 (1988) .
7. H.R. MacDonald, R. Schneider, R.K.Lees, R.C. Howe, H. Acha-Orbea, H. Festenstein, R.M. Zinkernagel, and H. Hengartner, T cell receptor Vβ use predicts reactivity and tolerance to Mls[a]-encoded antigens, *Nature* 332:40 (1988).
8. J.L. Roberts, S.O. Sharrow, and A. Singer, Clonal deletion and clonal anergy in the thymus induced by cellular elements with different radiation sensitivities, *J. Exp. Med.* 171:935 (1990) .
9. A. Kojima, and R.T. Prehn, Genetic susceptibility to post-thymectomy autoimmune diseases in mice, *Immunogenetics* 14:15 (1981) .
10. H. Smith, I.-M. Chen, R. Kubo, and K.S.K.Tung, Neonatal thymectomy results in a repertoire enriched in T cell deleted in adult thymus, *Science* 245:749 (1989) .
11. E.J. Yunis, R. Hong, M.A. Grewe, C. Martinez, E. Cornelius, and R.A. Good, Postthymectomy wasting associated with autoimmune phenomena. Antiglobulin-positive anemia in A and C57BL/6 Ks mice, *J. Exp. Med.* 125:947 (1967).
12. Frontiers in Research : Immunological Tolerance. *Science* 248, 1335-1393 (1990) .
13. W.N. Frankel, C. Rudy, J.M. Coffin, and B.T. Huber, Linkage of *Mls* genes to endogenous mammary tumour viruses in inbred mice, *Nature* 349: 526 (1991).

14. H. Acha-Orbea, A.N. Shakhov, L. Scarpellino, E. Kolb, V. Müller, A. Vessaz-Shaw, F. Fuchs, K Blöchlinger, P. Rollini, J. Billotte, M. Sarafidou, H.R. MacDonald, and H. Diggelmann, Clonal deletion of Vβ14-bearing T cells in mice transgenic for mammary tumour virus, *Nature* 350:207 (1991).

15. D.L. Woodland, M.P. Happ, K.J. Gollob, and E. Palmer, An endogenous retrovirus mediating deletion of αβ T cells? *Nature* 349:529 (1991).

16. D.J. Dyson, A.M. Knight, S. Fairchild, E. Simpson, and K. Tomonari, Genes encoding ligands for deletion of Vβ11 T cells cosegregate with mammary tumour virus genomes, *Nature* 349:531 (1991) .

17. C.Y. Okada, B. Holzmann, C. Guido, E. Palmer, and I.L.Weissman, Characterization of a rat monoclonal antibody specific for a determinant encoded by the Vβ7 gene segment. Depletion of Vβ7[+] T cells in mice with Mls-1[a] haplotype, *J. Immunol.* 144:3473 (1990) .

18. M.P. Happ, D.L. Woodland, and E. Palmer, A third T-cell receptor β-chain variable region gene encodes reactivity to *Mls-1*[a] gene products, Proc. Natl. Acad. Sci. USA 86: 6293 (1989) .

19. A.M. Pullen, P. Marrack, and J.W. Kappler, The T-cell repertoire is heavily influenced by tolerance to polymorphic self-antigens, *Nature* 335:796 (1988).

20. D. Woodland, M.P. Happ, J. Bill, and E. Palmer, Requirement for cotolerogenic gene products in the clonal deletion of I-E reactive T cells, *Science* 2 47:964 (1990).

21. J. Bill, O.Kanagawa, D.L. Woodland, and E. Palmer, The MHC molecule I-E is necessary but not sufficient for the clonal deletion of Vβ11-bearing T cells, *J. Exp. Med.* 169:1405 (1989).

22. M.S. Vacchio, and R.J. Hodes, Selective decreases in T cell receptor Vβ expression. Decreased expression of specific Vβ families is associated with expression of multiple MHC and non-MHC gene products, *J. Exp. Med.* 170:1335 (1989).

23. P. Kisielow, H. Blüthmann, U.D. Staerz, M. Steinmetz, H. von Boehmer, Tolerance in T-cell-receptor transgenic mice involves deletion of nonmature CD4[+]8[+] thymocytes, *Nature* 333:742 (1988).

24. K.M. Murphy, A.B. Heimberger, and D.Y. Loh, Induction of antigen of intrathymic apoptosis of CD4[+]CD8[+] TCR[lo] thymocytes *in vivo*, *Science* 250:1720 (1990).

25. M.K. Jenkins, and R.H. Schwartz, Antigen presentation by chemically modified splenocytes induces antigen-specific T cell unresponsiveness *in vitro* and *in vivo*, J. *Exp. Med.* 165:302 (1987).

26. M.K. Jenkins, P.S. Taylor, S.D. Norton, and K.B. Urdahl, CD28 delivers a costimulatory signal involved in antigen-specific IL-2 production by human T cells, *J. Immunol..* 147:2461 (1991).

27. D.L. Mueller, M.K. Jenkins and R.H. Schwartz, Clonal expansion versus functional clonal inactivation. A costimulatory signalling pathway determines the outcome of T cell antigen receptor occupancy, *Ann. Rev. Immunol.* 7:445 (1989).

28. C.H. June, J.A. Ledbetter, P.S. Linsley, and C.B. Thompson, Role of the CD28 receptor in the T cell activation, *Immunol. Today* 11:211 (1990).

29. P.S. Linsley, W. Brady, L. Grosmaire, A. Aruffo, N.K. Damle, and J.A. Ledbetter, Binding of the B cell activation antigen B7 to CD28 costimulates T cell proliferation and interleukin 2 mRNA accumulation, *J. Exp. Med.* 173;721 (1991)

30. M. Azuma, M. Cayabyab, D. Buck, J.H. Phillips, and L.L. Lanier, CD28 interaction with B7 costimulates primary allogeneic proliferative responses and cytotoxicity mediated by small, resting T lymphocytes, *J. Exp. Med.* 175:353 (1992).

31. L. Koulova, E.A. Clark, G. Shu, and B. Dupont, The CD28 ligand B7/BB1 provides costimulatory signal for alloreactivation of the CD4+ T cells, *J. Exp. Med.* 173:759 (1991).

32. C. Kim, K.A. Siminovitch, and A. Ochi, Reduction of lupus nephritis in MRL/*lpr* mice by a bacterial superantigen treatment, *J. Exp. Med.* 174:1431 (1991).

33. S.-H. Rhim, S.E. Millar, F. Robey, A.M. Luo, Y.-H. Lou, T. Yule, P. Allen, J. Dean, and K.S.K. Tung, Autoimmune disease of the ovary induced by a ZP3 peptide from the mouse zona pellucida, *J. Clin. Invest.* 89:28 (1992).

34. J.P. Clayton, G.M. Gammon , D.G. Ando, K.H. Kono, L. Hood, and E.E. Sercarz, Peptide-specific prevention of experimental allergic encephalomyelitis. Neonatal tolerance induced to the dominant T cell determinant of myelin basic protein, *J. Exp. Med.* 169:1681 (1989).

35. G.M. Gammon, A. Oki , N. Shastri, and E.E. Sercarz, Induction of tolerance to one determinant on a synthetic peptide does not affect the response to a second linked determinant. Implications for the mechanism of neonatl tolerance induction, *J. Exp. Med.* 164:667 (1986).

36. G. Gammon, K. Junn, N. Shastri, A. Oki, S. Wilbur, and E.E. Sercarz, (1986) Neonatal T-cell tolerance to minimal immunogenic peptides is caused by clonal inactivation, *Nature* 319:413 (1986).

37. J.M. Korostoff, M.T. Nakada, S.J. Faas, K.J. Blank, and G.N. Gaulton, Neonatal exposure to thymotropic Gross murine leukemia virus induces virus-specific immunologic nonresponsiveness, J. *Exp. Med.* 172:1765 (1990).

38. R. Michalides , R. van Nie, and R. Nusse, Mammary tumor induction loci in GR and DBAf mice contain one provirus of the mouse mammary tumor virus, *Cell* 23:165 (1981).

39. W.M. Kast, L. Roux, J. Curren, H.J.J. Blom, A.C. Voordouw, R.H. Meloen, D. Kolakofsky, and C.J.M. Melief, Protection against lethal Sendai virus infection by *in vivo* priming of virus-specific cytotoxic T lymphocytes with a free synthetic peptide, *Proc. Natl. Acad. Sci. USA* 88:2283 (1991).

40. T.N.M. Schumacher, M.L.H. de Bruijn, L.N., Vernie, W.M. Kast, C.J.M. Melief, J.J. Neefjes, and H.L. Ploegh, Peptide selection by MHC class I molecules, *Nature* 350:703 (1991).

41. T.N.M. Schumacher, M.-T. Heemels, J.J. Neefjes, W.M. Kast, C.J.M. Melief, and H.L. Ploegh, Direct binding of peptide to empty MHC class I moelcules on intact cells and *in vitro. Cell* 62:563 (1990).

42. M.L.H. de Bruijn, T.N.M. Schumacher, J.D. Nieland, H.L. Ploegh, W.M. Kast, and C.J.M. Melief, Peptide loading of empty MHC molecules on RMA-s cells allows induction of primary cytotoxic T lymphocytes responses, *Eur. J. Immunol.* 21:2963 (1991).

43. R.-H. Lin, M.J. Mamula, J.A. Hardin, and C.A. Janeway Jr, Induction of auto-reactive B cells allows priming of autoreactive T cells, *J. Exp. Med.* 173:1433 (1991).

44. F. Healy, C. Drouet, P. Romero, C. Jaulin, and J.L. Maryanski, Conversion of a self peptide sequence into a K^d-restricted neo-antigen by a Tyr substitution, *J. Exp. Med.* 174:1657 (1991).

45. P.S. Ohashi, S. Oehen, K. Buerki, H. Pircher, C.T. Ohashi, B. Odermatt, B. Malissen, R.M. Zinkernagel, and H. Hengartner, Ablation of "tolerance" and induction of diabetes by virus infection in viral antigen transgenic mice, *Cell* 65:305 (1991).

46. M.B.A. Oldstone, M. Nerenberg, P. Southern, J. Price, and H. Lewicki, Virus infection triggers insulin-dependent diabetes mellitus in a transgenic model: role of anti-self (virus) immune response, *Cell* 65:319 (1991).

47. C.J.P. Boog, J. Boes, and C.J.M. Melief, Stimulation with dendritic cells decreases or obviates the CD4+ helper cell requirement in cytotoxic T lymphocyte responses, *Eur. J. Immunol.* 18:219 (1988).

48. J.J. Neefjes, "Cell Biological Aspects of MHC Class I and II Molecules", Thesis, University of Amsterdam, The Netherlands, Rodopi, Amsterdam (1990).

49. M.L.H. de Bruijn, "Mechanisms of antigen presentation to T lymphocytes, Thesis, University of Leiden, The Netherlands (1992).

B-LYMPHOCYTE LINEAGE-COMMITTED, IL-7 AND STROMA CELL- REACTIVE PROGENITORS AND PRECURSORS, AND THEIR DIFFERENTIATION TO B CELLS

Fritz Melchers, Dirk Haasner, Martin Streb
and Antonius Rolink

Basel Institute for Immunology
Grenzacherstrasse 487
Postfach
4005 Basel, Switzerland

INTRODUCTION

Long term lymphoid cell cultures from fetal liver and bone marrow[1] have been used to study T- and B-lymphocyte development 'in vitro' and by repopulation experiments of suitable hosts 'in vivo'[2]. Cell contacts between the progenitors of lymphocytes and stromal cells, as well as cytokines produced upon these contacts[3] were found to be required for lymphopoiesis, specifically of the B-lymphocyte lineage[4-6]. Multiple contacts between stromal cells and progenitors have since been identified in this adhesion[7-11]. These early forms of progenitor and precursor cultures set the stage and raised the hope to dissect the hierarchy of hemopoietic-lymphopoietic differentiation to clone the differently committed cells, and to determine their capacity to proliferate and to differentiate.

PROGENITORS AND PRECURSORS OF THE B-LYMPHOCYTE LINEAGE

Culture conditions have now been established, and sources of progenitors and precursors found which allow the establishment of lines and clones of B lineage-committed cells with extensive pro-liferation capacity 'in vitro', and with capacities of differentiation along the B-lineage pathway 'in vitro' as well as 'in vivo'[12-17]. The cell lines and clones established in our laboratory[17] from fetal liver were shown to have the following characteristics. They proliferated with division times of a day in serum-substituted cultures under the stimulatory influence of adherent stromal cells and the cytokine IL-7 for periods longer than half a year. These lines expressed varying levels of

Mechanisms of Lymphocyte Activation and Immune Regulation IV: Cellular Communications
Edited by S. Gupta and T.A. Waldmann, Plenum Press, New York, 1992

111

the B lymphocyte lineage related markers PB76, B220, BP-1, V_{preB} and $\lambda 5$, but no surface Ig or MHC class II molecules. All clones expressed V_{preB} and $\lambda 5$ in a high percentage of cells, while B220 and/or BP-1 expression was low or undetectable in some. A cell line, and several clones established from it, had k and l light chain genes in germ-line configuration. Either one or both of the H-chain-gene containing chromosomes carried V_H to D_H and V_L to J_L rearrangements. This resulted in the development of varying percentages of surface Ig-positive, MHC class II negative LPS-reactive B cells within 2-3 days, in the absence of contacts with stromal cells and/or IL-7. When injected into SCID mice, the cultured pre B cells populated the spleen of these mice to 5% with surface Ig-, MHC class II-positive, LPS-reactive cells for >25 weeks. The long term 'in vitro' proliferative capacity of these D_H-J_H rearranged pre B cell clones makes them major candidates for committed stem cells of the B lineage.

THE C-KIT-ENCODED TYROSINE KINASE REGULATES THE PROLIFERATION OF EARLY PRE B CELLS

The proto-oncogene c-kit encodes a surface membrane bound member of the tyrosine kinase receptor family within the w (white spotting) locus in the mouse genome[18-22]. Mutations in this locus develop abnormalities in neural crest-derived melanocytes, primordial germ cells and hemopoietic stem cells[23].

Mice with mutation in the steel locus (sl) are phenotypically similar to w mutants in their pleiotropic dysfunctions in the development of melanocytes, primordial germ cells and hemopoietic stem cells. While the c-kit-encoded mutants of the w locus are intrinsic to the cells of the melanocyte, germ cell and hemopoietic cell lineages, sl mutants are defective in the microenvironment associated with the migration, homing and cell contact-dependent development of these lineages. A gene within the sl locus has recently been found to encode a growth factor for stem cells (steel factor; SLF), which acts on a wide variety erythroid, myeloid and lymphoid precursor cells and synergizes with granulocyte-monocyte (GM)-CSF in erythroid/myeloid development[24]. Purified SLF binds selectively to the c-kit-encoded tyrosine kinase receptor expressed on erythroid/myeloid precursor cells.

A monoclonal antibody (mAb; ACK2), recognizing the extracellular domains of the c-kit tyrosine kinase[25] has been employed to demonstrate that c-kit is involved in B lymphocyte development[26]. The c-kit-encoded tyrosine kinase is expressed on the surface of normal $D_H J_H$ rearranged murine pre B cell clones. These clones are inhibited by the mAb in their proliferation while remaining capable of differentiation to surface immunoglobulin-positive B cells. Stimulation of mature B cells by mitogens is unimpaired by the mAb. This indicates that c-kit regulates early antigen-independent, but not late antigen-dependent B cell development, that SLF may be a cytokine for pre B cells, and that c-kit may be an oncogene of the B lineage.

CHANGES IN FREQUENCES OF CLONABLE PRE B CELLS DURING LIFE IN DIFFERENT LYMPHOID ORGANS OF MICE

The high efficiency of plating on stromal cells in the presence of IL-7 of $D_H J_H$-rearranged precursor B cell lines and clones has suggested that the initial precursor cells residing in B cell generating organs could be cloned 'ex vivo'.

Limiting dilution analyses of cell populations from fetal liver, blood, spleen and bone marrow at different times before and after birth of a mouse have now indicated that the limiting cell in such analyses is the precursor B cell (Rolink, A., Haasner, D., Streb, M. and Melchers, F.; in preparation). The number of these cells in the various sites change during lifetime. A wave occurs before and until birth in liver. Up to two weeks after birth, high frequencies (1 in 100) of clonable pre B cells are present in the spleen. Thereafter their numbers in spleen constantly drop and become undetectable at 6 to 8 weeks of age. High frequencies (1 in 50) of clonable pre B cells are present in bone marrow up to 5 weeks of age. Thereafter they decrease to 1 in 2000-3000 at two years. While the expression of the pre B cell specific gene $V_{pre}B$ in liver and spleen correlates well with the clonable pre B cell frequencies in these organs at different stages during life, it does not do so in the bone marrow. $V_{pre}B$ expression in bone marrow decreases only approximately four fold from 4 weeks to 3 months of life. A similar reduction is observed in the number of B220-positive, surface IgM-negative pre B cells, while the frequencies of $D_H J_H$-rearranging or rearranged, clonable pre B cells in bone marrow decreases much more, i.e. from approximately 1 in 50 at 4 weeks to 1 in 1000 at 3 months in life. If the production of B cells is relatively constant throughout the life of a mouse, these results suggest that B cells may be generated at later times of life from $V_{pre}B/\lambda 5$-expressing precursors, which are not clonable on stromal cells and IL-7, and which might not have the same long-term proliferative capacity of the precursors seen in earlier life.

EARLY STAGES OF IL-7-REACTIVE, STROMAL CELL-DEPENDENT B LINEAGE PROGENITORS

The earliest progenitors proliferating on stromal cells in the presence of IL-7 have been isolated from young bone marrow of mice in which their RAG-2 gene is nonfunctional as a consequence of targeted integration of a mutated form of the gene[27]. These progenitors have all H and L chain loci in germline configuration, but already express $V_{pre}B$ and $\lambda 5$. These results are also supported by experiments with progenitors and precursors from early stages of B cell development in normal mice. Rearrangements in pre B cell clones early in B cell development from fetal liver and bone marrow appear to begin with nonrearranged H chain loci, indicating again that the earliest progenitors reactive to stromal cells and IL-7 have at least part of their Ig genes in germline configuration. Rearrangements in fetal liver do not insert N-sequences at the $D_H J_H$ joints. Rearrangements in spleen and bone marrow throughout life occur with N-regions inserted at $D_H J_H$ joints. At least half of all primary pre B cell clones develop mitogen-reactive B cells after differentiation to sIg^+ B cells (Rolink, A., Haasner, D., Streb, M. and Melchers, F.; in preparation).

EXPRESSION OF A μ HEAVY CHAIN ON THE SURFACE OF PRE B CELLS INTERFERES WITH THEIR CAPACITY TO PROLIFERATE ON STROMAL CELLS IN THE PRESENCE OF IL-7

Frequencies of precursor B cells in fetal liver and in bone marrow from μH chain-transgenic mice are at least 20 to100 fold lower than those of normal, nontransgenic littermates (Melchers, F., Bosma, G., Bosma, M. and Rolink, A.;

in preparation). This indicates that expression of μH chain on the surface of pre B cells, most likely associated with surrogate L chain $V_{pre}B/\lambda_5$ at that stage of development, terminates their capacity to proliferate on stromal cells in the presence of IL-7[ref.6]. It is likely that it allows normal differentiation to sIg+ B cells, involving the rearrangement of L chain loci, since the μH chain-trans-genic animals have sIg+ B cells in their peripheral lymphoid organs. Again, this could suggest that precursor B cells, in this case maybe cells expressing the $V_HD_HJ_H$-rearranged transgene, not reactive to stroma cells and IL-7, are generating the daily supply of sIg+-B cells.

DIFFERENTIATION TO $V_HD_HJ_H$-REARRANGED, μ H CHAIN AND V_LJ_L-REARRANGED, L-CHAIN EXPRESSING B CELLS

When IL-7 is removed from cultures of D_HJ_H-rearranged pre B cell clones they differentiate to sIg+ (as well as sIg-) B cells which can be stimulated by the polyclonal activator of lipopolysaccharide (LPS) to IgM-secreting cells. With three such pre B cell clones between 7% and 12% of IgM+ B cells could be detected in these differentiating cultures[28]. The majority (78-92%) of the IgM+ B cells co-expressed κ light chains. The percentage of λ light chain expressing B cells was below detectable level. Upon LPS stimulation, the percentages of IgM+ B cells increased dramatically (from 32-64%). The majority (91-97%) of the IgM+ B cells express κ chains, but a very small percentage (3.1-5.0%) express λ. A similarly high κ/λ ratio was found in 418 hybridomas prepared from these LPS-stimulated B cells (388 κ+ and 30 λ+). Thus, the high κ/λ ratio characteristic of the mouse peripheral B cell reportoire[29-33] is already evident in the antigen-independent transition from pre B to B cells.

The repertoires of V_H gene expression in newly generated B cells was determined by RNA dot-blot analysis with probes from 11 V_H families (Streb, M., Melchers, F. and Rolink, A.; in preparation). Hybridomas obtained by fusion with the myeloma cell X63 from LPS-stimulated sIg+ B cells, generated either 'in vivo' in SCID mice populated with these pre B cell clones, or 'in vitro' by differentiation of their three pre B cell clones in the absence of IL-7, were analyzed.

Relative to the number of V_H genes in a given V_H family, the most 3' V_H gene families (7183 and Q52) were 2 to 3 fold overrepresented, while the largest V_H gene family (J558) was approximately 2 fold underrepresented, and the 3609 V_H gene family was represented at normal frquencies. Five other V_H gene families were strongly underexpressed. Differentiation to sIg+, mitogen-reactive B cells also occur 'in vivo' when these pre B cell clones are injected in SCID mice, where they populate some of the B cell lineage compartments at least for several months. The expressed V_H repertoire was analyzed in 29 hybridomas generated from LPS-stimulated spleen cells of SCID mice populated with pre B cell clones. Within this collection of hybridomas, again the most 3' V_H gene families were overexpressed so that half expressed a member of the V_H 7183 family and one third a member of the V_H Q52 family. Of all the hybridomas generated from 'in vitro' or 'in vivo' differentiated B cells, approximately 40% were found to produce antibodies that bind to at least one of a panel of autoantigens. About 20% of these hybridomas produced multi-reactive autoantibodies, binding to more than one of the tested self antigens. The V_H gene repertoire of the autoantibody producing hybridomas is indistinguishable from the V_H gene repertoire used by the total collection of hybridomas. In conclusion, the B cell repertoires generated from $D_H J_H$-rearranged pre B cell clones 'in vitro' or 'in vivo' resemble the early B cell repertoire of fetal liver, neonatal or regenerating bone marrow. In the absence of T cells, no randomization of the repertoire is observed 'in vivo'.

CONCLUSIONS

The picture of B cell development which emerges from these studies is simple. In fetal liver and bone marrow, all available sites of stroma become populated during embryonic development with progenitors and precursors which have their IgL loci in germline, and their IgH loci either in germline or in $D_H J_H$-rearranged forms. In these states they are capable of long-term proliferation when transferred into secondary hosts, or when put into tissue culture on stromal cells in the presence of IL-7. Once all stromal cells are occupied any excess cells arising by division of progenitors and precursors will rearrange V_H to $D_H J_H$ and V_L to J_L, simply because they have lost contact to stroma and the neighbourhood of cells secreting IL-7. The early $D_H J_H$-rearranged, stroma cell/IL-7-reactive state is preserved by contact; differ-entiation, in this model, is single loss of contact; no new ligand appears to be needed to induce it. Since $Ig\mu H$ chain-transgenic mice produce sIg$^+$ B cells, and since old bone marrow with very few remaining $D_H J_H$-rearranged, stroma cell/IL-7-reactive precursor may continue to produce sIg$^+$ B cells, it may well be that a $V_H D_H J_H$-rearranged, L chain gene nonrearranged precursor population might continue to preserve pre B cells and generate sIg$^+$ B cells by unequal division. If B cell development is really continuous throughout life, this latter mode may suffice to generate the daily supply of new cells.

ACKNOWLEDGEMENTS

The Basel Institute for Immunology was founded and is supported by F. Hoffmann-La Roche Ltd., Basel, Switzerland.

REFERENCES

1. K.A. Denis and O.N. Witte, Long-term lymphoid cultures in the study of B cell differentiation, in: "Immunoglobulin Genes", T. Honjo, F.W. Alt and T.H. Rabbits, eds., Academic Press, New York (1989).
2. R.A. Phillips, Development and regulation of the lymphocyte lineages: an interpretative overview, Progr. Immunol. 7:305 (1989).
3. A.E. Namen, S. Lupton, K. Hjerrild, J. Wagnall, D.Y. Mochuzuki, A. Schmierer, B. Mosley, C. March, D. Urdal, S. Gillis, D. Cosman and R.G. Goodwin, Stimulation of B cell progenitors by cloned murine interleukin 7, Nature 333:571 (1988).
4. P.W. Kincade, G. Lee, C.E. Pietrangeli, S.I. Hayashi and J.M. Gimble, Cells and molecules that regulate B-lymphopoiesis in bone marrow, Ann. Rev. Immunol. 7:111 (1988).
5. K. Dorshkind, Regulation of hemopoiesis by bone marrow stromal cells and their products, Ann. Rev. Immunol. 8:111 (1990).

6. A. Rolink and F. Melchers, Molecular and cellular origins of B lymphocyte diversity, *Cell* 66:1081 (1991).
7. R.D. Sanderson, P. Lalor and M. Bernfield, B lymphocytes express and lose syndecan at specific stages of differentiation, *Cell Reg.* 1:27 (1989).
8. P.S. Thomas, C.E. Pietangeli, S.I. Hayashi, M. Schachner, C. Goridis, M. Low and P.W. Kincade, Demonstration of neural cell adhesion molecules on stromal cells which support lymphopoiesis, *Leukemia* 2:171 (1988).
9. P. Bernardi, V.P. Patel and H.F. Lodish, Lymphoid precursor cells adhere to two different sites on fibronectin. *J. Cell. Biol.* 105:489 (1988).
10. K. Miyake, C.B. Underhill, J. Lesley and P.W. Kincade, Hyaluronate can function as a cell adhesion molecule and CD44 participates in hyaluronate recognition. *J. Exp. Med.* 172:69 (1990).
11. K. Miyake, I.L. Weissman, J.S. Greenberger and P.W. Kincade, Evidence for a role of the integrin VLA-4 in lymphohemapoiesis. *J. Exp. Med.* 173:599 (1991).
12. L.S. Collins and K. Dorshkind, A stromal cell line from myeloid long-term bone marrow cultures can support myelopoiesis and B lymphopoiesis. *J. Immunol.* 138:1082 (1987).
13. C.A. Whitlock, G.F. Tidmarsh, C. Müller-Sieburg and I.L. Weissman, Bone marrow stomal cell lines with lymphopoietic activity express high levels of a pre-B neoplasia-associated molecule. *Cell* 48:1009 (1987).
14. S.I. Nishikawa, M. Ogawa, S. Nishikawa, T. Kumisada and H. Kodama, B lymphopoiesis on stromal cell clones: stromal cell clones acting on different stages of B cell differentiation. *Eur. J. Immunol.* 18:1767 (1988).
15. T. Era, M. Ogawa, S.I. Nishikawa, M. Okamoto, T. Honjo, K. Akagi, J.I. Migazaki and K.I. Yamamura, Differention of growth signal requirement of B lymphocyte precursor is directed by expression of immunoglobulin. *EMBO J.* 10:337 (1991).
16. S.I. Hayashi, T. Kunisada, M. Ogawa, T. Sudo, H. Kodama, T. Suda, T. Nishikawa and S.I. Nishikawa, Stepwise progression of B lineage differentiation supported by interleukin 7 and other stromal cell molecules. *J. Exp. Med.* 171:1683 (1990).
17. A. Rolink, A. Kudo, H. Karasuyama, Y. Kikuchi and F. Melchers, Long-term proliferating early pre-B cell lines and clones with the potential to develop to surface Ig-positive, mitogen-reactive B cells in vitro and in vivo. *EMBO J.* 10:327 (1991).
18. P. Besmer, J.E. Murphy, J.P.C. George, F. Qui, P.J. Bergold, L. Lederman and H.W. Snyder Jr., A new acute transforming feine retrovirus and relationship of its oncogene v-*kit* with the protein kinase family. *Nature* 320:415 (1986).
19. B. Chabot, D.A. Stephenson, V.M. Chapman, P. Besmer and A. Bernstein, The proto-oncogene *c-kit* encoding a transmembrane tyrosine kinase receptor maps to the mouse *W* locus. *Nature* 335:88 (1988).
20. E.N. Geissler, M.A. Ryan and D.E. Housman, The dominant-white spotting (*W*) locus of the mouse encodes the *c-kit* proto-oncogene. *Cell* 55:185 (1988).
21. J.C. Tau, K. Nocka, P. Ray, P. Traktman and P. Besmer, The dominant W[42] *spotting* phenotype results from a missense mutation in the *c-kit* receptor kinase. *Science* 247:209 (1990).
22. K. Nocka, J. Tau, E. Chiu, T.Y. Chu, P. Ray, P. Traktman and P. Besmer, Molecular basis of dominant negative and loss of function mutations at the murine *c-kit*/white spotting locus: W[37], W[v], W[41] and W. *EMBO J.* 9:1805 (1990).

23. D. Bennett, Developmental analysis of a mutation with pleiotropic effects in the mouse. *J. Morphol* 98:199 (1956).

24. O.W. Witte, Steel locus defines new multipotent growth factors. *Cell* 63:5-6 (1990).

25. M. Ogawa, Y. Matsuzaki, S. Nishikawa, S.I. Hayashi, T. Kunisada, T. Sudo, T. Kina, H. Nakauchi and S.I. Nishikawa, Expression and function of *c-kit* in hemopoietic progenitor cells. *J. Exp. Med.* 174:63 (1991).

26. A. Rolink, M. Streb, S.I. Nishikawa and F. Melchers, The c-kit-encoded tryrosine kinase regulates the proliferation of early pre-B cells, *Eur. J. Immunol.* 21:2609 (1991).

27. Y. Shinkai, G. Rathbun, K.-P. Lam, E.M. Oltz, V. Stewart, M. Mendelsohn, J. Charon, M. Datta, F. Young, A.M. Stall and F.W. Alt, RAG-2 deficient mice lack mature lymphocytes due to inability to initiate VDJ rearrangement. *Cell* in press.

28. A. Rolink, M. Streb and F. Melchers, The κ/λ ratio in surface immunoglobulin molecules on B lymphocytes differentiating from D_HJ_H-rearranged murine pre-B cell clones in vitro, *Eur. J. Immunol.* 21:2895 (1991).

29. K.L. Maguire and E.S. Vietta, κ/λ shifts do not occur during maturation of murine B cells. *J. Immunol.* 127:1670 (1981).

30. S. Kessler, K.J. Kim and I. Scher, Surface membrane κ and λ light chain expression on spleen cells of neonatal and maturing normal and immune-defective CBA/N mice: The $\kappa:\lambda$ ratio is constant. *J. Immunol.* 127:1674.(1981)

31. T. Takemori and K. Rajewsky, Lambda chain expression at different stages of ontogeny in C57BL/6, BALB/c and SJL mice. *Eur. J. Immunol.* 11:618 (1981).

32. J.M. LeJeune, D.E. Briles, A.R. Lawton and J.F. Kearney, Estimate of the light chain repertoire size of fetal and adult BALB/cJ and CBA/J mice. *J. Immunol.* 129:673.(1982)

33. H. Sauter and C.J. Paige, Detection of normal B-cell precursors that give rise to colonies producing both κ and λ light immunoglobulin chains. *Proc. Natl. Acad. Sci. USA* 84:4898 (1987).

REGULATORY CELLS AND CYTOKINES INVOLVED IN

PRIMARY B LYMPHOCYTE PRODUCTION

Kenneth Dorshkind, Ramaswamy Narayanan, and
Kenneth S. Landreth

Division of Biomedical Sciences
University of California, Riverside, CA 92521
Department of Molecular Genetics
Hoffmann-LaRoche, Inc., Nutley, NJ 07110,
and The Department of Microbiology and Immunology and
Mary Babb Randolph Cancer Center
West Virginia University, Morgantown, WV 26506

INTRODUCTION

Considerable progress has recently been made in defining cells and molecules that regulate the production of B lymphocytes in the bone marrow. Acquisition of this information is a direct result of advances in molecular biology and the development of cellular systems that permit the process of primary B cell differentiation to be duplicated in culture. As a result, it is possible now to present a working model of B cell development which identifies distinct proliferative and differentiative factors and the developmental stages on which they act.

BONE MARROW STROMAL CELLS SUPPORT DISTINCT STAGES OF B CELL DEVELOPMENT

It is now accepted that fibroblastoid cells, collectively referred to as stromal cells, present in the intersinusoidal spaces of the medullary cavity are a primary source of signals that regulate B lymphocyte development[1-3]. Characterization of these supporting cells was facilitated by the description of a long-term bone marrow culture system by Whitlock and Witte that allows stromal cell dependent B lymphopoiesis to occur in vitro[4]. These cultures are dependent upon establishment of an adherent layer of marrow derived cells which support the growth of B lineage cells at various stages of differentiation. Working on the assumption that the fibroblastoid cells present in these adherent layers were representative of stromal cells in situ, several groups cloned adherent cell populations with the aim of identifying those which could support B cell production. Several stromal cell lines were isolated, and their characterization indicated they exhibit a heterogeneity in their ability to support different stages of B cell differentiation.

One group of stromal cells can be defined based on their ability to support the development of pre-B cells. During this process, immature hemopoietic precursors develop into B cell progenitors/pre-B cells[5-9], identified by the expression of the 220,000 molecular weight B220 cell surface antigen and cytoplasmic immunoglobulin (Ig) μ heavy chain protein. One example of a stromal cell line that mediates this

Mechanisms of Lymphocyte Activation and Immune Regulation IV: Cellular Communications
Edited by S. Gupta and T.A. Waldmann, Plenum Press, New York, 1992

119

relatively early step in B lymphopoiesis is S17[5-7]. These stromal cells, or medium conditioned by them, allow pre-B cells to be generated from fresh bone marrow depleted of B220 and cμ expressing populations. Whether or not the target cell on which the S17 cells act is already committed to B lymphopoiesis is not clear.

While the S17 cell line supported development of pre-B cells, their further maturation into immunoglobulin light chain protein expressing cells does not occur. This indicates that once progenitors have developed to the pre-B cell stage, they still require additional stimuli to mature to surface Ig expression. Working on the assumption that a detailed characterization of these pre-B cells would provide information about the precise developmental stage at which such signals act, the status of light chain gene rearrangements and expression in the stromal cell dependent pre-B cells was analyzed. The cells were obtained from lymphoid colonies whose growth in semisolid medium was potentiated by factors from S17 stromal cells[7]. The results of these studies indicated that kappa light chain genes in many of these pre-B cells were rearranged, but no mature kappa transcripts could be detected.

Other stromal cell lines, such as S10[7] and SCL 160[10], provide the conditions under which pre-B cells can express light chain protein and mature into B lymphocytes. These observations indicate that stromal cells must be included among the accessory cell populations demonstrated to mediate pre-B cell maturation[11] and provide additional evidence that the pre-B to B cell transition is a regulated event.

STROMAL CELL DERIVED SOLUBLE MEDIATORS

The above results provide strong evidence that stromal cell signals act at defined stages during B cell development but do not provide a molecular basis for these observations. However, the recent cloning of genes that encode stromal cell derived cytokines has made it possible to identify specific mediators involved in B cell development. Two factors on which considerable attention has focused in this regard are IL-7[12] and kit-Ligand (KL)[13]. IL-7 was identified by its ability to stimulate the proliferation of B220[+] pre-B cells. There is no described effect of KL alone on B lineage cells, but it exhibits potent synergy with IL-7 in potentiating pre-B cell growth[14,15]. With the description of these and other mediators, it is appropriate to ask which mediate the developmental events supported by the different stromal cell lines.

Development of Pre-B Cells From Immature Hemopoietic Precursors

As noted, the S17 stromal cell line is able to support the generation of B220/cμ expressing pre-B cells from hemopoietic precursors that do not express these determinants. Using a short-term liquid culture assay system in which pre-B cells develop from their precursors, it was shown that no defined interleukin or colony stimulating factor could account for the S17 activity. In particular, neither IL-7 or KL supported this developmental step. This latter observation was consistent with the fact that no IL-7 message could be detected in S17 cells by Northern analysis or following thirty replication cycles of the polymerase chain reaction[14].

Further evidence that this stromal cell activity is not KL, IL-7 or any other defined interleukin or colony stimulating activity was provided by biochemical analyses indicating that the apparent molecular weight of the active molecule(s) in S17 conditioned medium was just under 10 kD[6]. The size of most defined interleukins and colony stimulating factors is 15 kD and above. Considerable efforts are underway to identify this low molecular weight S17 stromal cell derived mediator which stimulates pre-B cell formation.

Proliferation of B lineage Cells

The short term assay system employed above revealed that only a minimal amount of cell proliferation occurred during the generation of pre-B cells from their precursors. However, since up to 5×10^7 B cells are generated each day in the bone marrow[16], additional signals that stimulate cell growth must also exist.

During the course of characterizing the S17 differentiation activity, conditioned medium from the line was observed to synergise with IL-7 in potentiating the growth of

B lineage cells. Since this effect could be replaced with recombinant KL, this suggested that factor was responsible for the synergistic activity and that IL-7 and KL together played an important role in expanding the numbers of B lineage cells. Further characterization of the KL and IL-7 effect indicated that these cytokines have no effect on the growth of precursors prior to their development into pre-B cells. Rather, they only appear to affect the proliferation of cells once they have developed into progenitors/pre-B cells in response to the S17 activity[14].

KL and IL-7 do not act on Ig expressing lymphocytes, since IL-7 does not stimulate proliferation of mature B cells[17] and antibodies to c-kit, the KL receptor, do not interfere with mitogen stimulated proliferation of splenocytes[18]. These results suggest that at some point during the course of $V_L J_L$ gene rearrangements and expression, receptors for IL-7 and KL are no longer expressed[17-19]. Analysis of pre-B cells present in individual stromal cell dependent lymphoid colonies indicated that by the time light chain genes are rearranged in pre-B cells, mRNA for c-kit and the IL-7 receptor are no longer expressed. Taken together, this minimizes a role for either KL or IL-7 in mediating the pre-B to B cell transition.

Development of B Cells From Pre-B Cells

Evidence was provided above that selected stromal cell lines provide conditions under which pre-B cells mature into surface Ig expressing B lymphocytes. Although the factors that mediate this event are not completely defined, IL-4 has been reported to stimulate surface Ig expression on pre-B cells in vitro[10,20].

SUMMARY

Based upon the above data, it is now possible to formulate a working model that defines the stages of B cell development on which stromal cells and their products act. During the initial stages of this process, pro-B cells which do not express Ig heavy or light chain protein or other non-Ig B lineage associated molecules develop into B220 and $c\mu$ expressing pre-B cells in response to a low (<10 kD) molecular weight stromal cell derived factor. No defined interleukin or colony stimulating factor, including molecules such as KL and IL-7, can replace stromal cell conditioned medium in mediating this developmental step.

There appears to be little cell proliferation associated with the differentiation of pro-B cells into pre-B cells. However, our data indicate that as precursors develop into B220 expressing B cell progenitors, they become sensitive to the proliferation stimulating effects of IL-7 and KL. These results are in accord with findings that progenitor cells that have undergone DJ_H rearrangements are particularly sensitive to KL and IL-7[18,19].

The analysis of pre-B cells present in individual lymphoid colonies indicates that once cells have rearranged and expressed their Ig heavy chain genes, they are no longer sensitive to KL and IL-7. These observations are based on the fact that receptors for these cytokines are not expressed in stromal cell dependent pre-B cells and are consistent with kinetic studies showing that the maturation of pre-B cells into surface Ig expressing B lymphocytes is not dependent upon cell proliferation[21]. Finally, pre-B cells mature into surface immunoglobulin expressing B lymphocytes, and stromal cells can provide the conditions under which this maturational step occurs. An IL-4-like molecule has been identified in this process[10], but what, if any, additional mediators might also be involved is not clear. Regardless of their identity, it will be of interest to determine if these stromal cell signals activate transcription factors that regulate Ig light chain protein expression.

This working model identifies the developmental stages on which different stromal cell derived signals act to potentiate steady state B lymphocyte production in the bone marrow, but it by no means excludes a role for other stromal and non-stromal cell derived signals. In view of the redundancy in many cytokine pathways, it is reasonable to assume that additional mediators will be described that affect B lymphopoiesis. It will be equally important in this regard to investigate cytokines which act as negative regulators as well.

REFERENCES

1. K. Dorshkind, Regulation of hemopoiesis by bone marrow stromal cells and their products, Annu. Rev. Immunol. 8:111 (1990).

2. P.W. Kincade, G. Lee, C.E. Pietrangeli, S.I. Hayashi, and J.M. Gimble, Cells and molecules that regulate B lymphopoiesis in bone marrow, Annu. Rev. Immunol. 7:111 (1989).

3. K. Dorshkind and K.S. Landreth, Regulation of B cell differentiation by bone marrow stromal cells, Int. J. Cell Cloning 10:in press (1992).

4. C.A. Whitlock and O.N. Witte, Long-term culture of B lymphocytes and their precursors from murine bone marrow, Proc. Natl. Acad. Sci. USA 79:3608 (1982).

5. L.S. Collins and K. Dorshkind, A stromal cell line from myeloid long-term bone marrow cultures can support myelopoiesis and B lymphopoiesis, J. Immunol. 138:1082 (1987).

6. K.S. Landreth and K. Dorshkind, Pre-B cell generation potentiated by soluble factors from a bone marrow stromal cell line, J. Immunol. 140:845 (1988).

7. A.J. Henderson, A. Johnson, and K. Dorshkind, Functional characterization of two stromal cell lines that support B lymphopoiesis, J. Immunol. 145:423 (1990).

8. C.A. Whitlock, G.F. Tidmarsh, C. Muller-Sieburg, and I.L. Weissman, Bone marrow stromal cell lines with high levels of a pre-B neoplasia-associated molecule, Cell 48:1009 (1987).

9. S.I. Nishikawa, M. Ogawa, S. Nishikawa, T. Kunisada, and H. Kodama, B lymphopoiesis on stromal cell clone: stromal cell clones acting on different stages of B cell differentiation, Eur. J. Immunol. 18:1767 (1988).

10. A.G. King, D. Wierda, and K.S. Landreth, Bone marrow stromal cell regulation of B-lymphopoiesis I. The role of macrophages, IL-1, and IL-4 in pre-B cell maturation, J. Immunol. 141:2016 (1988).

11. P.W. Kincade, G. Lee, C.J. Paige, and M.P. Schied, Cellular interactions affecting the maturation of murine B lymphocyte precursors in vitro, J. Immunol. 127:255 (1981).

12. A.E. Namen, S. Lupton, K. Hjerrild, J. Wagnall, D.Y. Mochizuki, A. Schmierer, B. Mosley, C. March, D. Urdal, S. Gillis, D. Cosman, and R.G. Goodwin, Stimulation of B-cell progenitors by cloned murine interleukin-7, Nature 333:571 (1988).

13. O.N. Witte, Steel locus defines new multipotent growth factor, Cell 63:5 (1990).

14. L.G. Billips, D. Petitte, K. Dorshkind, R. Narayanan, C.P. Chiu, and K.S. Landreth, Differential roles of stromal cells, interleukin-7, and kit-ligand in the regulation of B lymphopoiesis, Blood 79:in press (1992).

15. I.K. McNeice, K.E. Langley, and K.M. Zsebo, The role of recombinant stem cell factor in early B cell development, J. Immunol. 146:3785 (1991).

16. D.G. Opsteltin and D.G. Osmond, Pre-B cells in the bone marrow: immunofluorescence stathmokinetic studies of the proliferation of cytoplasmic μ-chain bearing cells in normal mice, J. Immunol. 131:2635 (1983).

17. G. Lee, A.E. Namen, S. Gillis, L.R. Ellingsworth, and P.W. Kincade, Normal B cell precursors responsive to recombinant murine IL-7 and inhibition of IL-7 activity by transforming growth factor-ß, J. Immunol. 142:3875 (1989).

18. A. Rolink, M. Streb, S.I. Nishikawa, and F. Melchers, The c-kit encoded tyrosine kinase regulates the proliferation of early pre-B cells, Eur. J. Immunol. 21:2609 (1991).

19. R.R. Hardy, C.E. Carmack, S.A. Shinton, J.D. Kemp, and K. Hayakawa, Resolution and characterization of pro-B and pre-pro-B cell stages in normal mouse bone marrow, J. Exp. Med. 173:1213 (1991).

20. A. Simons and D. Zhary, The role of IL-4 in the generation of B lymphocytes in the bone marrow, J. Immunol. 143:2540 (1989).

21. K.S. Landreth, C. Rosse, and J. Clagett, Myelogenous production and maturation of B lymphocytes in the mouse, J. Immunol. 127:2027 (1981).

THE ROLE OF IL-7 AND ITS RECEPTOR
IN B-CELL ONTOGENY

Linda S. Park[1], Philip J. Morrissey[2],
Barry Davison[3], and Kenneth Grabstein[2]

[1]Department of Biochemistry
[2]Department of Immunology
[3]Department of Transgenics
Immunex Research and Development Corporation
Seattle, Washington 98101

INTRODUCTION

Interleukin-7 is a molecule that was originally identified on the basis of its ability to stimulate growth of murine pre-B cells (1). It appears to be principally a growth factor, not a differentiation factor, and has no effects on mature B cells. How early IL-7 acts in the pathway of B-lymphopoiesis has not been clearly defined; however, it is clear that other factors or signals are required at the very earliest stages of B-lymphopoiesis due to the necessity for contact with a stromal layer. In order to better understand the role of IL-7 in the context of the other factors which may influence B-lymphopoiesis, we have studied the effect of altering *in vivo* levels of IL-7 in mice. This has been done in three ways: first, by relatively short term administration of IL-7; second, through the generation of IL-7 transgenic mice which result in constitutive overexpression of IL-7; and third, through administration to mice of an anti-IL-7 antibody to remove the *in vivo* effects of IL-7. Although IL-7 has known effects on a number of cell lineages, the data presented here will concentrate primarily on what happens to the B-cell compartment.

RESULTS AND DISCUSSION

In Vivo Administration of IL-7

C57BL/6 mice (10-12 weeks of age) were injected subcutaneously with IL-7, and the cellularity of lymph node, spleen and bone marrow was examined beginning on day 1 after cessation of a six-day treatment with IL-7. There was a significant increase in the cellularity of the spleen, bone marrow and mesenteric lymph node on day 1 post treatment. On day 3 post treatment, spleen and lymph node cellularity were still high, while bone marrow was down to background; by day 9 all levels were down to background. Thymic cellularity did not differ from control mice at any point. Looking at the peripheral organs, spleen and lymph node in animals

Mechanisms of Lymphocyte Activation and Immune Regulation IV: Cellular Communications
Edited by S. Gupta and T.A. Waldmann, Plenum Press, New York, 1992

125

receiving IL-7 showed a significant increase in B220$^+$/sIgM$^+$ cells which gradually dropped back down to baseline within a few days after cessation of IL-7 administration. Because IL-7 has no direct effects on mature B-cells *in vitro*, these increases are presumably the result of effects of IL-7 on the pre-B cell compartment. The effects of IL-7 treatment on B220$^+$/sIgM$^-$ pre-B cells in bone marrow and spleen were also examined. In both cases, a significant increase in the number of pre-B cells was observed which was maximal shortly after the cessation of IL-7 treatment, dropping back down to baseline within a few days following cessation of IL-7 administration. We used a three-color FACS analysis to further look at the expanding cell population in the bone marrow. Cells harvested one day after the end of a six-day course of IL-7 were examined for BP-1 and B220 expression when gated on sIgM$^-$ cells. The percentage of cells staining positive for B220 and BP-1 was dramatically increased upon IL-7 treatment, while the BP-1$^-$ population was relatively unaffected. These data could suggest that the BP-1$^+$ population is the earliest population to respond to IL-7, while the B220$^+$/BP-1$^-$ population, which is less mature, is unresponsive.

The increase in pre-B cells in the bone marrow and spleen is accompanied by a similar increase in mature B220$^+$/sIgM$^+$ cells in both spleen and lymph node. Because mature B cells do not respond directly to IL-7, these cells are likely to be the differentiated progeny of the expanded pre-B cell compartment. The increase in mature cells is greatest immediately following cessation of IL-7 treatment and gradually falls close to baseline within a few days. The relatively rapid fall off would either indicate that the life-span of mature B cells in these organs is normally relatively short, or, perhaps alternatively, that faced with an oversupply of mature B cells, some regulatory mechanism kicks in to bring the numbers down to more normal levels.

IL-7 Transgenic Mice

In comparison to mice receiving short courses of IL-7 treatment, what happens to mice exposed to continuously elevated levels of IL-7? To examine this question, IL-7 transgenic mice were generated with the IL-7 gene hooked to either the Metallothionine promoter or Lck promoter. The metallothionine promoter results in non-cell-type specific expression primarily in liver, kidney and gut, as well as other organs which results in high circulating serum levels of IL-7. This promoter is not active in lymphoid cells. In contrast, the Lck promoter directs essentially T cell specific expression in both thymocytes and mature T cells. Only very low level expression occurs in B cells, and serum levels are low.

The Lck/IL-7 transgenic mice showed virtually no perturbation in the B cell compartment. Within a few months, however, most of these animals die of complications associated with severely enlarged mesenteric lymph nodes. Surface antigen analysis of this lymph node hyperplasia shows that the majority of cells are Thy-1$^+$/CD3$^-$ and are either CD8$^+$/CD4$^-$ or CD4$^+$/CD8$^+$. Thus, they resemble early thymocytes. These cells are tumorigenic when transplanted into nude mice again resulting in extreme lymph node hyperplasia. Studies of just what these cells are and what is happening in these mice continue.

In contrast to these Lck/IL-7 mice, constitutive expression of IL-7 off the Metallothionine promoter produced a very benign phenotype. The cellularity of lymphoid organs, spleen, lymph node, thymus and bone marrow, were examined, and all showed increased cellularity in the MT/IL-7 mice. The increase was particularly dramatic in spleen and lymph node where cellularity was increased several hundred percent. When cell phenotypes in the spleen were examined, we found that the B cell compartment, B220$^+$ and sIgM+ cells, were increased relative to control cells, while CD4+ and CD8+ cells were decreased. Overall, the composition of the spleen had shifted toward a higher percentage of B lymphoid cells relative to T lymphoid cells. In the bone marrow, B220+ cells in MT/IL-7 mice were markedly increased relative to control mice, apparently to a small degree at the expense of the myeloid lineage as indicated by a decrease in MAC-1+ positive cells. These analyses were done at about 5 months of age; however, a number of the MT/IL-7 mice lived out a normal life span without developing tumors or any other life-threatening conditions. It appears from these mice that the presence of constitutively higher levels of IL-7 has kept B-lymphopoiesis operating at a higher rate than normal; however, this has not resulted in any chronic ailments or life-threatening conditions over the normal life span of the animal.

In Vivo administration of anti-IL-7 antibody

What happens if IL-7 is taken away instead of increased *in vivo* in mice? This question was approached through the use of an anti-IL-7 antibody called M25 which is a murine IgG2b generated against human IL-7. This antibody has the ability to neutralize both human and murine IL-7. M25 can completely block proliferation of the IL-7-dependent cell line IxN2b in response to either human or murine IL-7; however, it is a little more effective against human IL-7. This is not surprising since it was generated using the human IL-7 molecule as an immunogen. Another way to look at the effectiveness of M25 is to examine its effects on Whitlock-Witte cultures (2). Normal Whitlock-Witte cultures contain clusters of lymphoid cells associated with an adherent stromal cell layer. When M25 is added to such a culture, the stromal layer remains intact, while virtually all the lymphoid cells are eliminated. These results are seen within 48 hours of adding M25 to an established culture. If M25 is added at the beginning of the culture period, the stromal cell layer will develop as usual, but no lymphoid cells will appear.

To test the effects of M25 *in vivo*, mice received 3 mg of M25, or an isotype matched control on days 0, 3, 6, 9. Then on day 10, bone marrow was harvested and the number of colony forming cells enumerated in a seven-day semi-solid agar bone marrow assay. We observed that CFU-GM were unchanged by M25 treatment, while the number of CFU-IL-7 were dramatically reduced. Different M25 treatment regimens were examined by looking at the decrease in the CFU-IL-7, and the nine-day treatment with 3 mg every three days was found to be sufficient for a maximal decrease in CFU-IL-7. It is also important to note that the number of CFU-IL-7 in these M25 treated mice was not zero, and increased length of treatment or increased amounts of M25 could not reduce it to zero. This may indicate that there is a population of IL-7 responsive cells which is being spared from M25 treatment.

To look at the effects of M25 in more depth, the published work of R.R. Hardy *et al.* (3) was used as a model. In this work, B220+/IgM- B-lineage cells in mouse bone marrow were resolved into four fractions based on differential expression of four cell surface antigens, and these subsets correlated with Ig gene rearrangement status, apparent requirement of these cells for stromal contact, and the presence of IL-7. The temporal expression of these antigens was used as a model in a series of experiments in which bone marrow cells isolated from M25 and control treated mice were examined by two- and three-color FACS analysis. The antigens which were examined are B220 (CD45R) which is present from pre-pro-B cells to mature B cells; BP-1, which is found on late pro-B cells, but not early cells; HSA or Heat stable antigen which is maximal at the pre-B stage and declines at earlier stages; and S7 which is a rat monoclonal against leukosialin (CD43) and is present in the earliest pro-B cells Hardy measured. IgM and B220 expression were first examined in control and M25 treated mice. M25 treatment was found to eliminate most of the B220+/IgM- population as well as the B220+/IgM+ population in the bone marrow. A very bright B220+/IgM+ population was unaffected. The B cell population gated on B220+ cells was then examined for IgM versus IgD expression. In the normal mice there is a population of B220+/IgM- cells, a population of IgM+/IgD- cells, and a population of IgM+/IgD+ cells. In the mast treated animals, all the IgM+/IgD- cells were eliminated but not the IgM+/IgD+ population. There was also a large reduction in the B220+/IgM- population, although some of these cells have been spared. Given that most of the pre and immature B cells have been eliminated, it is quite likely that these IgM+/IgD+ cells have recirculated to the bone marrow from the periphery.

We next examined what happens at the pre and pro B stages. Within the B cell compartment in the control animals, there are both BP-1- and BP-1+ populations, as well as a BP-1-/B220 bright population. In M25 treated mice the BP-1+ cells (which are pre B and late pro B) are eliminated. The BP-1- population has been reduced but not eliminated. To look more closely at the subsets within this BP-1- population, in order to narrow down the cells escaping M25 treatment, the HSA and S7 markers were examined. In control animals, there are B220^{low}, HSA^{hi} and B220^{hi}, HSA^{low} populations. In the M25 treated animals, the HSA^{hi} population is essentially entirely wiped out, and a remaining population of B220^{hi}, HSA^{low} cells is again primarily the mature IgD+ population. The final marker examined was S7, which is present on the earliest B-lineage cells in R.R. Hardy's model (3). Cells gated for B220+ were examined for S7 versus HSA expression. Within the earliest S7+ population in control animals, there is an HSA^{hi} and an HSA^{low} population where the HSA^{low} would be expected to be an earlier B-lineage population than the HSA^{hi}. In M25 treated animals, the S7+/HSA^{hi} population has been eliminated and the remaining population is the earliest S7+/HSA^{low} cells.

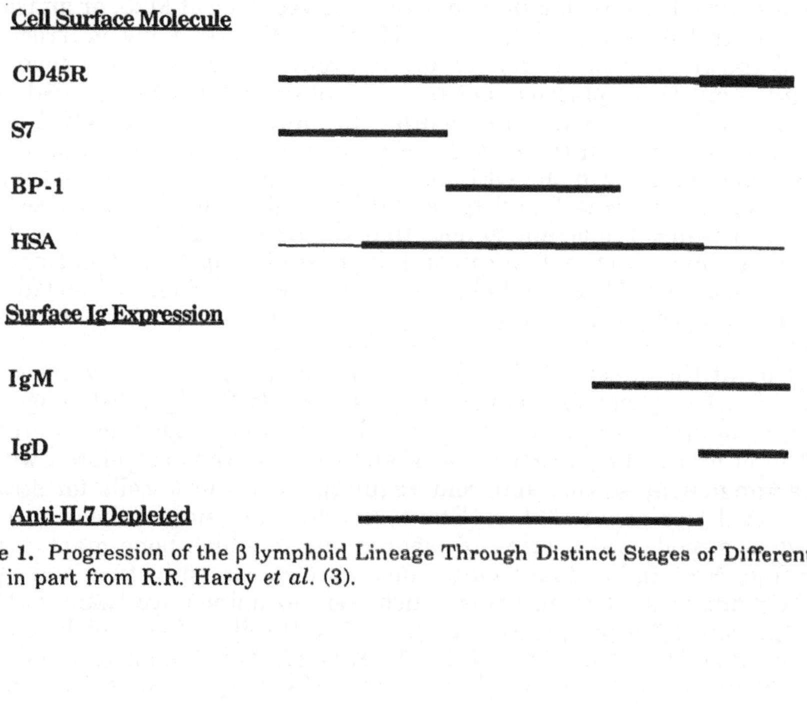

Cell Surface Molecule

CD45R ————————————————————————

S7 ——————————

BP-1 ——————————

HSA — ——————————————— —

Surface Ig Expression

IgM ———————————

IgD ——————

Anti-IL7 Depleted ————————————————————

Figure 1. Progression of the β lymphoid Lineage Through Distinct Stages of Differentiation. Taken in part from R.R. Hardy *et al.* (3).

Figure 1 summarizes these and other results from FACS analyses in a schematic covering the pre-pro B cell stage to the mature B cell stage. As shown in the figure, M25 eliminates all but the early pro-B cell stage in the bone marrow, and a mature IgM+/IgD+ population is also left which has probably resulted from mature cells migrating back into the bone marrow from the periphery. In contrast to the effects in the bone marrow, after a ten-day course of M25, only minor effects were seen in the peripheral organs. When the M25 treatment was extended for three to six weeks, with the animal continuing to receive 3 mg of M25 every 3 days, there was still minimal effect in the periphery. This is despite the fact that pro and pre B cells in the bone marrow are virtually wiped out within the first few days of treatment. While there was some decrease in B cells in the spleen after 6 weeks, it was still less than a 25% drop. Only a small drop in cellularity was seen in the thymus and, perhaps not surprisingly, there was no reduction in the CD5/Ly-1 cells in the peritoneal cavity. This very small reduction in mature B cells in the spleen would suggest that turnover of B cells in the periphery is actually very slow.

Returning to the original model of R.R. Hardy (3), the cells that were spared from M25 elimination were S7^{hi}/HSA^{low} which corresponds to pre-pro B cells. Hardy found these cells also had an absolute requirement for stromal contact. Are these cells really an IL-7 independent population? The fact that they are spared by M25 would suggest that they are; however, it is also possible that the absolute requirement for stromal contact may be indicative of an anatomical position in the bone marrow which may preclude entrance of the antibody, or some other kind of steric constraint, due to the close lymphocyte/stromal cell contact. As mentioned above, when the bone marrow of M25 treated animals is put into a 7-day CFU-IL-7 colony assay, there is a small but significant number of CFU-IL-7 colonies formed showing that IL-7 responsive cells do remain in the marrow. If the remaining B lineage cells from M25 treated animals are enriched by panning on B220 and put into a colony assay, it was found that this population was extremely responsive to IL-7 and formed very large colonies. While this does not prove these cells are IL-7 dependent, it does suggest that the earliest pre-pro B cells are highly responsive to IL-7.

Additional studies are required to define how early in the pathway of lymphopoiesis IL-7 is required or has effects. It is hoped that studies with mice in which either the IL-7 or IL-7R gene has been knocked out by homologous recombination will help provide a clearer picture of the complete *in vivo* role of IL-7 in lymphoid development.

REFERENCES

1. A.E. Namen, A.E. Schmierer, C.J. March, R.W. Overell, L.S. Park, D.L. Urdal, and D.Y. Mochizuki, B cell precursor growth-promoting activity: purification and characterization of a growth factor active on lymphocyte precursors, *J. Exp. Med.* 167:988 (1988).
2. C.A. Whitlock, D. Robertson, and O.N. Witte, Murine B cell lymphopoiesis in long term culture, *J. Immunol. Meth.* 67:353 (1984).
3. R.R. Hardy, C.E. Carmack, S.A. Shinton, J.D. Kemp, and K. Hayakawa, Resolution and characterization of Pro-B and Pre-Pro-B cell stages in normal mouse bone marrow, *J. Exp. Med.* 173:1213 (1991).

ROLE OF CONTACT AND SOLUBLE FACTORS IN THE GROWTH AND DIFFERENTIATION OF B CELLS BY HELPER T CELLS

Randolph J. Noelle[*][1], Lisa Marshall[*], Meenakshi Roy[*],
David M. Shepherd[*], Ivan Stamenkovic[^], Jeffrey A. Ledbetter[#],
Alejandro Aruffo[#] and H. Perry Fell[#]

[*]Department of Microbiology
Dartmouth Medical School
Lebanon, NH 03756
[#]Bristol-Myers Squibb Pharmaceutical Research Institute
Seattle, WA 98121
[^]Department of Pathology
Massachusetts General Hospital
Harvard Medical School
Boston, MA 02114

INTRODUCTION

Antigen-specific, class II-restricted interactions have been repeatedly shown to be essential in the initiation of humoral immunity. In addition to TCR, and CD4, other T_h molecules, like CD28, ICAM-1, LFA-1 and B7 subserve co-stimulatory functions (Fig. 1).

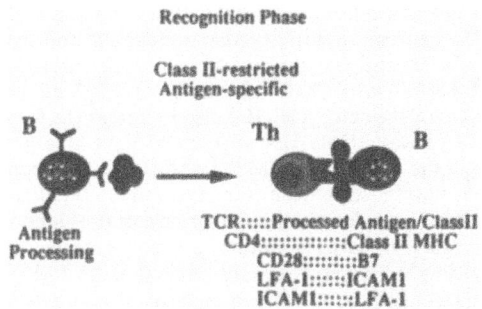

Figure 1. Recognition stage of cognate help.

Although these interactions are essential in the early stages of T_h-dependent B cell recognition and T_h activation, recent evidence indicates that the actual events that trigger B cell activation are class II-unrestricted and antigen-non-specific. Experiments indicate that metabolically inactivated T_h are capable of forming antigen-specific conjugates with B cells, yet do not induce B cell cycle entry. Therefore, class II-restricted, antigen-specific interactions

---

[1] To whom correspondence should be addressed.

Mechanisms of Lymphocyte Activation and Immune Regulation IV: Cellular Communications
Edited by S. Gupta and T.A. Waldmann, Plenum Press, New York, 1992

between T_h and B cells do not induce B cell cycle entry. It suggests that none of the resident T_h surface molecules (CD4, ICAM-1, CD5, etc.) mediate B lymphocyte activation. It is only *after T_h activation* that T_h can be metabolically poisoned and retain the ability to induce B cells to enter the cell cycle.[1] This function of T_h to induce B cell cycle entry is referred to as the *effector function*. The effector function is antigen-nonspecific and class II-unrestricted,[1-3] making it distinct from those restrictions that govern the initial interactions with antigen-presenting B cells. Therefore, after class II-restricted, antigen-specific interaction with B cells, T_h become activated to express membrane proteins that are capable of reciprocally activating B cells via non-restricted mechanism(s) (Fig 2.).

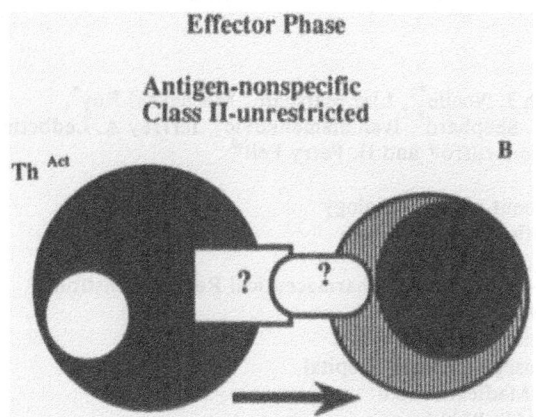

Figure 2. Effector phase of cognate help.

Once induced into the cell cycle by activated T_h, B cells become responsive to the growth and differentiative effects of lymphokines. The role of lymphokines in T_h-dependent B cell, growth and antibody synthesis has been difficult to resolve because of the endogenous synthesis of lymphokines by T_h and the uncertainty if neutralizing anti-lymphokine mabs can effectively eliminate lymphokine activity. Purified plasma membranes from activated T_h (PM^{ACT}) have been used to study the effector phase of cognate help. PM^{ACT} from activated T_h clones, but not resting clones (PM^{REST}) express an activity that induces B cell cycle entry and B cell lymphokine responsiveness in an antigen-nonspecific, class-II unrestricted manner. Using PM^{ACT}, it has been shown[4] that IL4 with PM^{ACT} from T_h1 and T_h2 increases the frequency of B cells that enter the G_{1a} phase of the cell cycle. IL4 also increases the frequency of B cells that progress into the G_{1b} and S phases. Only IL4 appears to have this growth factor activity. Although B cell proliferate in the presence of PM^{ACT} and IL4, no Ig secretion is observed. In the presence of IL4 *and* IL5, PM^{ACT} have been shown to induce the differentiation of B cells to produce all isotypes of Ig.[4,5] These data establish that signals provided by T_h PM and soluble factors provide all the necessary signals for inducing B cell growth, differentiation and Ig isotype switching. Finally, PM^{ACT} from T_h1 were equivalent to PM^{ACT} from T_h2 in their ability to induce B cell differentiation in the presence of IL4 and IL5. This conclusively dismissed the notion that T_h1 were poor "helpers" because they did not express appropriate membrane proteins to activate B cells. Therefore, in summary, IL4 is necessary and sufficient for T_h-dependent B cell growth. No other lymphokines appear to enhance growth induced by PM^{ACT} and IL4. Differentiation to Ig synthesis by B cells activated by T_h requires both IL4 *and* IL5 (Fig. 3).

MOLECULAR BASIS FOR THE EFFECTOR PHASE OF COGNATE HELP

Studies suggest that upon activation, T_h express a membrane protein that can induce the entry of B cells into the cell cycle. Antibodies specific to α,β TCR, CD3, ICAM-1, LFA-1,

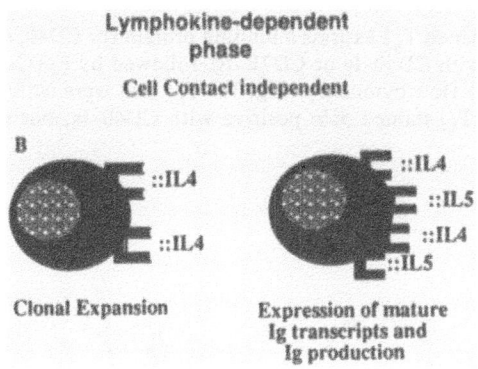

Lymphokine-dependent
phase

Cell Contact independent

Clonal Expansion | Expression of mature Ig transcripts and Ig production

Thy-1, CD4, CD3, CD2, MHC class I do not inhibit the ability of PM^{ACT} to activate B cells. Therefore, although some of these molecules may play an accessory role in T_h effector function, it does not appear that any of these molecules are essential for inducing B cell cycle entry.

In the human system, it has been shown that anti-CD40 mab induced B cell proliferation.[6] Such results implicated CD40 as an important triggering molecule for B cells. To determine if CD40 was involved in the induction of B cell RNA synthesis by PM^{ACT}, a soluble fusion protein of the extracellular domains of human CD40 and the Fc domain of human IgG_1 (CD40-Ig) was added to cultures of PM^{ACT} and B cells. PM^{ACT} derived from T_h1 were prepared and used to stimulate B cell RNA synthesis. The addition of CD40-Ig to culture caused a dose-dependent inhibition of B cell RNA synthesis that was induced by PM^{ACT} (Fig. 4). A CD7-Ig fusion protein[7] was without effect even when used at 25 µg/ml. CD40-Ig did not inhibit the activation of B cells by T-independent activators, like LPS (data not shown).

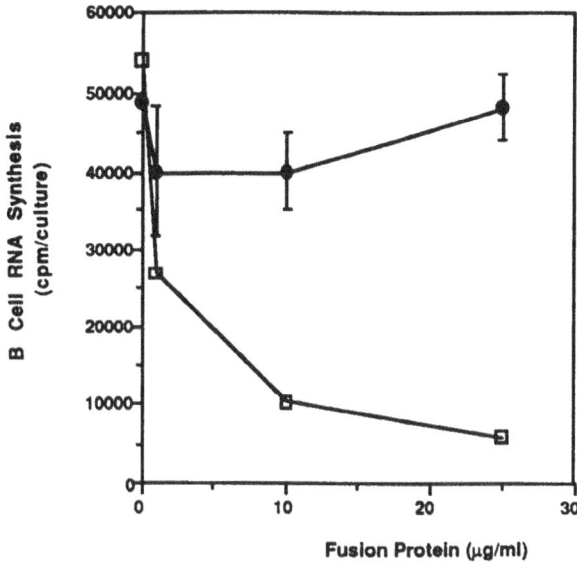

Figure 4. CD40-Ig inhibits B cell cycle entry. Resting B cells were cultured with PM^{ACT} from T_h1. To these cultures, no fusion protein was added or 1-25 µg/ml of CD40-Ig (□) and CD7E-Ig (●) was added. B cell RNA synthesis was assessed. Results presented are the arithmetic means of triplicate cultures +/- s.d., and are representative of 3 such experiments.

To investigate whether activated T_h1 express a binding protein for CD40, resting and activated (16 hrs) T_h1 were stained with CD40-Ig or CD7E-Ig, followed by FITC-anti-HIgG. Binding of CD40-Ig was assessed by flow cytometry (Fig. 3). T_h1 that were activated for 16 hrs with anti-CD3, but not resting T_h, stained 56% positive with CD40-Ig, but not with the control CD7E-Ig.

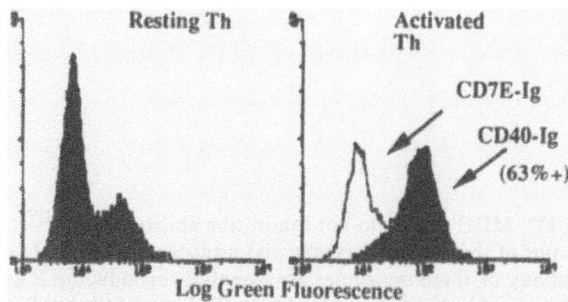

Figure 5. CD40-Ig detects a molecule expressed on activated, but not resting T_h. Resting and activated T_h were incubated with fusion proteins for 20 minutes at 4°C, followed by FITC-conjugated goat anti-hIgG (25 µg/ml). Percentage positive cells and MFI were determined by the analysis of at least 5000 cells/sample. Results are representative of 6 such experiments.

To identify the CD40-Ig binding protein, T_h1 proteins were biosynthetically labelled with ^{35}S methionine/cysteine and proteins immunoprecipitated with CD40-Ig or CD7E-Ig. The immunoprecipitated proteins were resolved by SDS-PAGE and fluorography. A prominent band with an apparent molecular weight of 39 kD immunoprecipitated with 10 µg of CD40/sample (data not shown). A 39 kd band was also immunoprecipitated from activated T_h that were vectorially labelled with ^{125}I, confirming that the 39kD protein was a membrane protein. The immunoprecipitated 39kD band was identical in size when resolved by SDS-PAGE under non-reducing conditions, indicating the CD40 binding protein was a single chain molecule. We have recently identified a hamster anti-mouse mab, MR1, specific to the 39kD T cell surface molecule. Like CD40-Ig, MR1 stained 50-60% of activated, but not resting T_h1 and blocked the activation of B cells by PM[ACT].

In conclusion, studies using polyclonally-activated T_h and T_h PM[ACT] demonstrate that a contact-dependent signal activates B cells in an antigen-nonspecific, genetically-unrestricted manner.[2,4,5] These findings indicate that non-polymorphic molecules on the surface of T_h and B cells mediated the induction of B cell cycle activation. CD40 was proposed to be an important ligand in T_h-dependent B cell activation since many functional responses of B cells to crosslinking of CD40 and to activated T_h were similar. For example, PM[ACT] and α-CD40 induce B cell cycle entry.[8-10] Further, α-CD40 and activated T_h in the presence of IL4, induced potent IgE production from resting B cells.[8,11-13] Finally, α-CD40 and PM[ACT] induced homotypic B cell aggregation.[14] Taken together with the data presented herein, the 39kD protein expressed by activated T_h and CD40 may represent the ligand-receptor pair essential for the induction of T_h-dependent B cell responses.

BIOCHEMICAL BASIS FOR THE EFFECTOR PHASE OF COGNATE HELP

Once B cells are activated, a number of different biochemical pathways are stimulated which transduce the signal across the plasma membrane to the interior of the cell. Regulators of B cell activation, including antibodies directed against membrane immunoglobulin, mIgM and mIgD, have traditionally been used to study the biochemical mechanisms of signal transduction in B cells. These studies have shown that crosslinking membrane immunoglobulin stimulates phosphoinositide hydrolysis resulting in calcium mobilization and the activation of protein kinase C. Additionally, crosslinking membrane immunoglobulin induces the phosphorylation of protein tyrosine residues, an event intimately tied with the regulation of cell growth. T_h-B cell interactions, which involve the ligation of several surface molecules, trigger B cell activation through biochemical events which differ from the anti-Ig pathway. Present studies indicate that unlike cross-linking surface immunoglobulin, T_h-B contact fails to: a) activate protein kinase C; b) mobilize calcium; c) generate cAMP; or d) stimulate gross changes in the tyrosine phosphorylation.

To gain insights into the possible signal transducing systems that were operative in T_h-dependent B cell cycle entry, B cells were induced to enter cycle with anti-Ig, LPS and PM^{ACT}. Inhibitors of specific second messenger systems were employed to evaluate the involvement of specific pathways in the signalling process (Fig. 6). Staurosporine and tyrphostin, inhibitors of PKC and an EGF-receptor related tyrosine kinase, respectively, inhibited B cell RNA synthesis induced by anti-sIg but not by LPS or Th-B interaction. In addition, forskolin, an activator of adenylate cyclase, inhibited both anti-sIg and LPS stimulated B cell RNA synthesis without affecting the Th-dependent B cell response. Finally, genistein, a specific inhibitor of tyrosine kinases, inhibited all induction of B cell RNA synthesis. Together, these data indicate that while LPS and Th-B induced B cell cycle entry share a similar biochemical pathway, with the exception of the role of cAMP, the anti-sIg and Th-B induced B cell responses signal through different biochemical pathways which share a common tyrosine kinase.

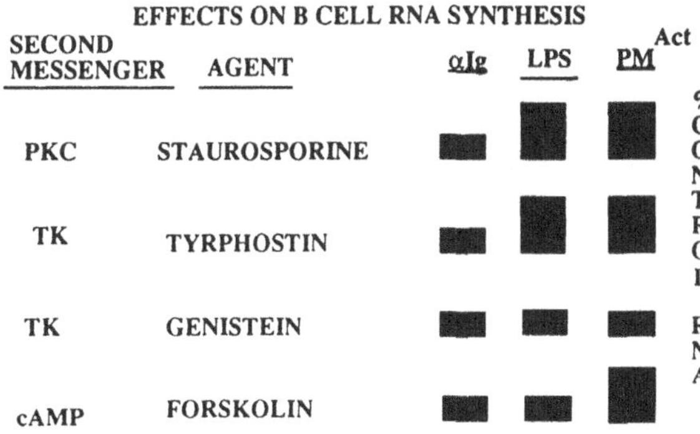

Figure 6. Involvement of second messenger cascades in B cell activation.

As stated previously, CD40 has been implicated as being an important ligand in Th-dependent B cell activation. Several CD40-regulated serine-threonine protein kinases have been identified. In addition, the CD40 receptor has been shown to be intimately associated with a protein tyrosine kinase/protein tyrosine phosphatase regulatory pathway.[15] Recent studies indicate that although there is no gross alteration in the phosphorylation pattern of phosphotyrosine proteins, T_h-dependent B cell activation is inhibited in the presence of genistein, a specific tyrosine kinase inhibitor. Further, ongoing studies suggest that ERK1 and ERK 2,[16] serine-threonine protein kinases, are phosphorylated on tyrosine residues following T_h-B cell interaction. Together, these data add further evidence that CD40 may part of the ligand-receptor pair essential for the induction of Th-dependent B cell responses.

LYMPHOKINE-DEPENDENT PHASE

Under conditions that support B cell growth (PM^{ACT} and IL4), no Ig secretion is observed. Secretion of IgM and IgG$_1$ requires PM^{ACT}, IL4 *and* IL5. This system demonstrates that IL5 is an essential factor for Ig production by conventional, resting B cells. It was possible that IL5 was exerting its positive effect on IgG$_1$ expression as a direct consequence of inducing germline transcripts from the γ_1 locus. Resting B cells have been shown to express germline γ_1 transcripts when stimulated with IL4.[17,18] It has also been reported that the inclusion of LPS heightened the level of germline γ_1 expression.[17] To determine whether lymphokines or PM^{ACT} in the presence or absence of lymphokines induced or altered the expression of germline γ_1 transcripts, a PCR-based technology was employed to analyze the relative abundance of specific Ig transcripts. Germline γ_1 transcripts were amplified using a 5' oligo specific for sequences that lie downstream of the majority of known germline initiation sites[19] but preceding the I splice donor site, and a 3' oligo specific for the γ_1 hinge region. Each RNA sample was titrated and the steady state level of β_2 microglobulin was assessed to assure uniformity in RNA isolation, reverse transcription and amplification.

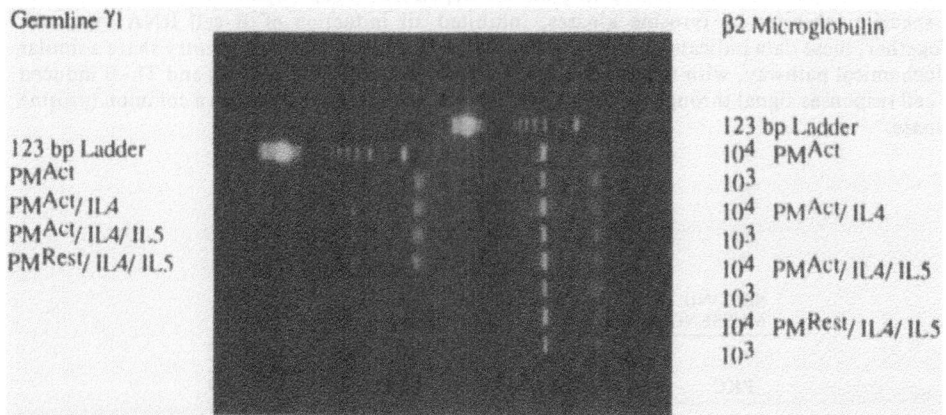

Figure 7. PM^{ACT} do not induce the expression of germline γ_1 transcripts. B cells (5x10^4/well) were cultured with PM^{ACT} or PM^{REST} (0.5 μg/well), IL4 (10 ng/ml) and/or IL5 (500 U/ml). On day 1, B cells were harvested, the number of viable cells ascertained and RNA extracted. PCR amplification (40 cycles) for the detection of germline γ_1 transcripts from 10^5 cell equivalents was performed. The expression of steady-state levels of β_2 microglobulin mRNA were ascertained by titration of input cDNA from 10^3-10^4 cell equivalents of cDNA. The experiment shown is representative of 3 such experiments.

Similar to published studies,[17,18,20] IL4 alone, but not LPS alone, induced the appearance of germline γ_1 transcripts after one day in culture (data not shown). Extended expression of germline γ_1 transcripts was observed in B cells stimulated with LPS/IL4 where germline γ_1 transcripts were abundantly expressed for up to four days. On day one, any condition containing IL4 (PM^{ACT}/IL4, PM^{ACT}/IL4/IL5) induced germline γ_1 transcripts similar to that observed with IL4 alone (data not shown). However, germline γ_1 transcripts endured for up to four days only when B cells were activated with PM^{ACT}/IL4 or PM^{ACT}/IL4/IL5.

The data thus far indicated that the inclusion of PM^{ACT} extended the duration in expression of germline γ_1 transcripts in the presence of IL4. Therefore, it was important to evaluate if PM^{ACT} alone could induce the appearance of germline γ_1 transcripts. When B cells were cultured with PM^{ACT} in the absence of IL4, no germline γ_1 transcripts were detected (Figure 7). As observed in previous experiments, the addition of IL4 (in the presence or absence of IL5) induced the appearance of germline γ_1 transcripts. PM^{REST} and PM^{ACT} in the presence of IL4/IL5 induced similar levels of germline γ_1 transcripts on day 1. This indicates that the levels of germline γ_1 transcript expression observed with IL4/IL5 is independent of activation status of the PM. β_2 microglobulin cDNA was amplified and titered from all samples. Upon amplification of limiting amounts of cDNA, subsaturating PCR products for β_2 microglobulin were generated. Although not well depicted in the figure, β_2 microglobulin PCR products were titred and similar in amount between groups. Minimal sample to sample variation was noted and indicated that equivalent amounts of cDNA were present in all groups.

Since the addition of IL5 to PM^{ACT}/IL4 induced the secretion of both IgM and IgG_1, without grossly altering the steady-state levels of germline γ_1 transcripts, the affects of IL5 on the levels of secreted μ and mature γ_1 transcripts were determined. No transcripts for secreted μ or γ_1 mature transcripts were detectable when B cells were cultured with PM^{ACT}/IL4. However, consistent with the secretion data, the combination of PM^{ACT}/IL4/IL5, LPS alone, or LPS/IL4 resulted in an increase in the steady-state levels of secreted μ and γ_1 mature transcripts (data not shown. In summary, contact of T_h and B cells did not result in the appearance of germline transcripts for γ_1. The requirement for IL5 in the production of IgM and IgG_1 appears to be due to its ability to induce the appearance of mature transcripts for μ and γ_1. A summary of these results is listed in Table 1.

Table 1. Role of lymphokines in the regulation of Ig expression.

	Protein		mRNA		
			Sterile	Mature	
Addition:	IgM	IgG1	γ_1	μ	γ_1
IL4	-	-	+	-	-
IL5	-	-	-	-	-
PM CD3	-	-	-	-	-
PM rest	-	-	-	-	-
PM CD3 IL4	-	-	+	-	-
PM rest IL4/IL5	-	-	+	-	-
PM CD3 IL4/IL5	+	+	+	+	+

ACKNOWLEDGEMENT

This work was supported in part by NIH grants AI26296, and GM37767.

REFERENCES

1. R.J. Noelle, J. McCann, L. Marshall, and W.C. Bartlett, Cognate interactions between helper T cells and B cells. III. Contact-dependent, lymphokine-independent induction of B cell cycle entry by activated helper T cells, *J. Immunol.*, 143, 1807 (1989).
2. S. Hirohata, D.F. Jelinek, and P.E. Lipsky, T cell-dependent activation of B cell proliferation and differentiation by immobilized monoclonal antibodies to CD3, *J. Immunol*, 140, 3736 (1988).
3. T. Owens, A noncognate interaction with anti-receptor antibody-activated helper T cells induces small resting murine B cells to proliferate and to secrete antibody, *Eur. J. Immunol*, 18, 395 (1988).
4. R.J. Noelle, J. Daum, W.C. Bartlett, J. McCann, and D.M. Shepherd, Cognate interactions between helper T cells and B cells. V. Reconstitution of T helper cell fuction using purified plasma membranes from activated Th1 and Th2 helper cells and lymphokines, *J. Immunol*, 146, 1118 (1991).

5. P.D. Hodgkin, L.C. Yamashita, R.L. Coffman, and M.R. Kehry, Separation of events mediating B cell proliferation and Ig production by using T cell membranes and lymphokines, *J. Immunol*, 145, 2025 (1990).

6. E.A. Clark and P.J.L. Lane, Regulation of human B cell activaiton and adhesion, *Annual Reviews of Immunol*, 9, 97 (1991).

7. N.K. Damle and A. Aruffo, Vascular cell adhesion molecule 1 induces T-cell antigen receptor-dependent activation of CD4+T lymphocytes, *Proc. Natl. Acad. Sci. USA*, 88, 6403 (1991).

8. F. Rousset, E. Garcia, and J. Banchereau, Cytokine-induced proliferation and immunoglobulin production of human B lymphocytes triggered through their CD40 antigen, *J. Exp. Med*, 173, 705 (1991).

9. J. Banchereau, P. de Paoli, A. Valle, E. Garcia, and F. Rousset, Long-term human B cell lines dependent on interleukin-4 and antibody to CD40, *Science* 251, 70 (1991).

10. P. Bjorck, B. Axelsson, and S. Paulie, Expression of CD40 and CD43 during activation of human B lymphocytes, *Scand. J. Immunol*, 33, 211 (1991).

11. K. Zhang, E.A. Clark, and A. Saxon, CD40 stimulation provides an IFN-gamma-independent and IL-4-dependent differentiation signal directly to human B cells for IgE production, *J. Immunol*, 146, 1836 (1991).

12. H. Gascan, J.F. Gauchat, G. Aversa, P. Van Vlasselaar, and J.E de Vries, Anti-CD40 monoclonal antibodies or CD4+ T cell clones and IL-4 induce IgG4 and IgE switching in purified human B cells via different signaling pathways, *J. Immunol*, 147, (1991).

13. H.H. Jabara, S.M. Fu, R.S. Geha, and D. Vercelli, CD40 and IgE: synergism between anti-CD40 monoclonal antibody and interleukin 4 in the induction of IgE synthesis by highly purified human B cells, *J. Exp. Med*, 172, 1861 (1990).

14. T.B. Barrett, G. Shu, and E.A. Clark, CD40 signaling activates CD11a/CD18 (LFA-1)-mediated adhesion in B cells, *J. Immunol*, 146, 1722 (1991).

15. F.M. Uckun, G.L. Schieven, I. Dibirdik, L.M. Chandan, A.L. Tuel, J.A. Ledbetter, Stimulation of protein tyrosine phosphorylation, phosphoinositide turnover, and multiple previously unidentified serine/threonine-specific protein kinases by the Pan-B-cell receptor CD40/Bp50 at discrete developmental stages of human B-cell ontogeny, *J. Biol. Chem*, 266:17478 (1991).

16. M.H. Cobb, D.J. Robbins, and T.G. Boulton, ERKs, extracellular signal-regulated MAP-2 kinases, *Curr. Opin. Cell. Biol*, 3, 1025 (1991).

17. M.T. Berton, J.W. Uhr, and E.S. Vitetta, Synthesis of germ-line gamma 1 immunoglobulin heavy-chain transcripts in resting B cells: induction by interleukin 4 and inhibition by interferon gamma, *Proc Natl Acad Sci U S A* 86, 2829 (1989).

18. C. Esser and A. Radbruch, Rapid induction of transcription of unrearranged sg1 switch regions in activated murine B cells by interleukin 4, *EMBO J*, 8, 483 (1989).

19. M. Xu and J. Stavnezer, Structure of germline immunoglobulin heavy chain γ_1 transcripts in interleukin 4 treated mouse spleen cells, *Dev. Immunol*, 1, 11 (1990).

20. E. Severinson, C. Fernandez, and J. Stavnezer, Induction of germ-line immunoglobulin heavy chain transcripts by mitogens and interleukins prior to switch recombination. *Eur. J. Immunol*, 20, 1079 (1990).

B-CELL ACTIVATION MEDIATED BY INTERACTIONS
WITH MEMBRANES FROM HELPER T CELLS

Marilyn R. Kehry,[1] Brian E. Castle,[1] and Philip D. Hodgkin[2]

[1]Department of Molecular Biology
 Boehringer Ingelheim Pharmaceuticals, Inc.
 Ridgefield, CT 06877
[2]The John Curtin School of Medical Research
 Australian National University
 Canberra City, A. C. T. Australia 2601

INTRODUCTION

Resting B lymphocytes are small dense cells that express cell surface IgM and IgD. Resting B cells from mouse spleen can be stimulated to proliferate and differentiate to immunoglobulin secretion by the combined action of direct contact with activated helper T (Th) cell clones and lymphokines secreted by Th cells. The mechanism by which this productive interaction of B cells and Th cells is mediated is most likely a multi-step process as outlined in Figure 1[1]. Specific B cells internal-

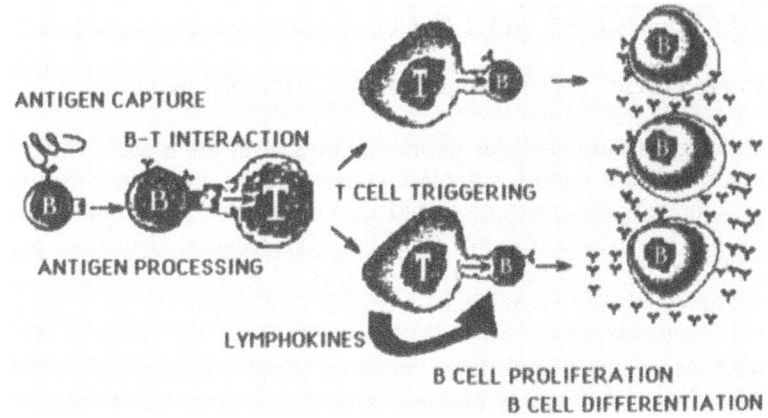

Figure 1. A model of Th cell-dependent B cell activation and differentiation. Adapted from (1).

ize protein antigens, degrade them to peptides and present the peptides on their cell surface in association with class II MHC molecules. This MHC-antigen complex on an antigen presenting B cell is recognized by an antigen-specific Th cell and results in Th cell activation. Surface molecules involved in this interaction are class II MHC, T cell receptor and associated proteins, and adhesion molecules such as LFA-1 and its ligands. Ligation of the T cell receptor complex induces the expression of a set of activation-specific genes that encode secreted lymphokines, proteins that modify existing molecules to potentially modulate their activity, and possibly new cell surface molecules. A reciprocal interaction of altered or newly synthesized Th cell surface effector molecules with ligands on the B cell surface then is involved in inducing B cell activation and proliferation. The molecules involved in this interaction have not yet been defined, but most likely consist of adhesion and effector molecules on the Th cell and receptor molecules, such as CD40, on the B cell surface. Subsequently, lymphokines produced by the activated Th cell enhance the proliferation of the B cell and effect differentiation, switching of immunoglobulin constant regions from μ to other isotypes, and immunoglobulin secretion.

Because the initial activation of the Th cell is MHC restricted and involves antigen-T cell receptor interaction, it has been difficult to define the requirements for the actual activation of the B cell once Th cell stimulation is complete. A series of studies using fixed activated Th cells or bystander B cells of a different MHC haplotype has shown that activated Th cells stimulate B cells in a cell contact-dependent, MHC-independent and antigen-independent manner.[2-5] Recently, we and others have shown that preparations of plasma membranes from activated Th cell clones are sufficient to reproduce the contact-dependent signals delivered by Th cells to B cells.[6-9] The Th membrane system provides a more defined and direct method of studying Th contact-dependent B cell activation and allows a more detailed investigation of the cell surface ligand-receptor interactions involved in productive Th cell-B cell contact. We briefly summarize below the properties and activities of Th cell membranes and describe what we have learned about B cell activation by studying this system. We will then focus on recent studies of two aspects of resting B cell stimulation by Th membranes: (1) Requirements for the induction of B cell proliferation, and (2) Requirements for the induction of antibody synthesis.

ACTIVATION OF B CELLS WITH Th CELL PLASMA MEMBRANES

Th Membranes Provide the Th-dependent Contact Signal to B Cells

Initial studies of enriched plasma membranes from the mouse Th2 cell clone D10 demonstrated that resting splenic B cells could be efficiently stimulated to proliferate by membranes from activated D10 cells but not by membranes from resting cells.[6] The Th2 membranes were not MHC-restricted in their action on dense B cells, and the addition of the D10-specific antigen, conalbumin, to the cultures had no effect on Th membrane-induced B cell proliferation. These studies showed that B cells responded specifically to cell contact signals provided by Th membranes.[6] Moreover, membranes prepared from activated Th1 and Th2 cell clones that synthesize different lymphokines seemed to provide similar signals to a majority of resting B cells.[7] Our studies were in agreement with bystander experiments that demonstrated that an activated Th cell was capable of stimulating MHC-nonidentical B cells in an antigen-independent fashion.[2,4,10]

Characteristics of the Active Component(s) in Th Membranes

Initial characterization of the active component(s) in the Th membranes that is interacting with a ligand(s) on the resting B cell has defined some of its properties but has not yet identified the component(s). Experiments using blocking antibodies to lymphokines in combination with membranes from Th1 or Th2 cell clones have excluded several known lymphokines, including IL-2, IL-3, IL-4, IL-5, IL-6, and IFN-gamma, from consideration as active components of the Th membranes.[6,7] Additionally, IL-4 has been shown to enhance the proliferative response of B cells to Th membranes, but is not itself required for the induction of proliferation.[6-8] The activity in the membranes was clearly associated with membrane protein, since digestion with several different proteases removed activity from the membranes.[6]

The expression of the activity in the Th membranes after Th cell activation also required mRNA and protein synthesis and was inhibited by cyclosporin A treatment of Th cells.[1] Additionally, the activity was expressed only transiently after Th cell activation and decreased gradually after 12 hours of Th cell activation.[1,6] These properties suggested that the activity was dependent on the expression of an activation-specific Th cell gene(s). Either the gene product(s) is expressed directly on the Th cell surface to alter Th cell function or is involved in modifying and activating an existing Th cell surface molecule.

A series of studies using antibodies to various membrane proteins has also demonstrated that Thy-1, CD23, B220 (CD45R), CD4, and ICAM-1 are not involved in the signaling interactions between Th membranes and B cells.[1] Additionally, we have shown that the regions of class II MHC molecules that interact with CD4 and the T cell antigen receptor during antigen presentation are not involved in reciprocal signaling of the B cell in this system.[6]

INDUCTION OF B CELL PROLIFERATION

Induction of B Cell Proliferation is Cell Density Independent

The Th membrane system was utilized to further characterize the requirements of B cells for entering into the cell cycle. Over a 2,000-fold range of B cell dose, the number of cells synthesizing DNA as a result of stimulation with Th membranes was directly proportional to input B cell number (Figure 2). This feature was similar to B cell stimulation by intact Th cells and antigen and is very different from the density-dependent activation of B cells by either LPS or anti-IgM (Figure 2). This dose-response suggests that the interaction of Th membranes and B cells provides all the requisite signals for activating and driving B cells through the cell cycle without the need for intercellular B cell-B cell contact or secretion of a soluble factor.

The Requirement for Extended Th Membrane-B Cell Contact

The minimum length of time of Th membrane-B cell contact required for the induction of B cell proliferation was examined by washing out Th membranes after various times of culture. Even after a 24 hour contact period, only 20% of the maximal proliferative response was seen at 48 hours. A 42 hour contact period only generated 70% of the maximal response. Thus, extended contact of B cells and Th membranes was required to induce maximal B cell growth. This suggests

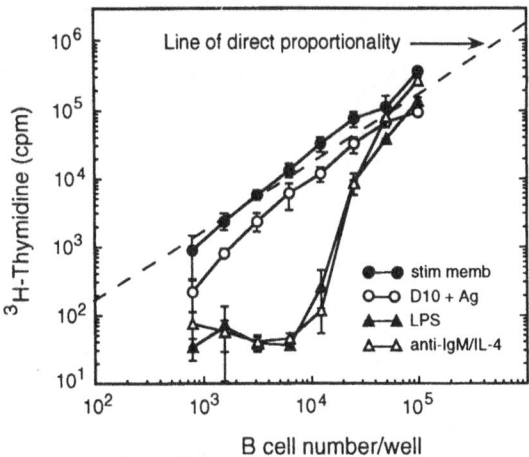

Figure 2. Relationship between B cell dose and proliferation. Dense B cells were titrated and cultured with four different stimuli: Goat anti-mouse IgM and 100 U/ml IL-4 (open triangles); LPS at 100 µg/ml (closed triangles); irradiated D10 Th2 cells and 100 µg/ml conalbumin as antigen (open circles); or membranes from activated D10 Th2 cells (closed circles). Cultures were incubated for 48 hours with a 4-hour pulse of [^{3}H]Thymidine. Each point is the mean ± standard error. Reprinted by permission from (1).

that resting B cells require continuous signaling over the first 24-36 hours to (overcome a threshold and) achieve induction of DNA synthesis.

Th membrane-B cell contact is, however, an early event in the *in vitro* cultures. This was demonstrated by staining, after various times of coculture, either B cells with anti-B220 or Th membranes with anti-CD4. Fluorescence microscopy demonstrated that direct contact of the Th membranes with all B cells occurred by 24 hours of culture. Additionally, aggregation of the B cells in very large clusters was induced and found to be maximal by 72 hours (data not shown). These responses are similar to the activation of adhesion in human B cells that have been stimulated through surface CD40 molecules.[11] Preparations of membrane vesicles from activated Th cells were also found to be extensively aggregated (see next section), and these extensively agglutinated Th membranes were readily observed bound to B cells after 24 hours of culture. However, during the course of culturing with B cells (over 4 days), Th membranes seemed to deaggregate and appeared as patchy bright dots on the surfaces of B cells (data not shown). Thus, the homotypic Th cell adhesion that initially is critical in inducing crosslinking of the B cell surface (see below) is reduced or eliminated as B cells enlarge and divide. Although the Th membrane vesicles deaggregate, contact between the Th membranes and B cells is maintained.

B Cell Surface Crosslinking and the Induction of Proliferation

To determine the physical character of active Th membrane vesicles, at various time during preparation and after sonication Th cell membranes were tested for their ability to stimulate dense B cell proliferation. Transmission electron microscopy was performed on each sample in parallel with the biological assays. Th membrane vesicles freshly prepared by homogenization of activated Th cells and centrifugation onto a sucrose pad contained little activity (Figure 3). These vesicles ranged in size between 10 and 1500 nm with some small aggregates up to 2 μm in size (data not shown). After ultracentrifugation these vesicles had maximal activity in stimulating B cell proliferation (Figure 3) and contained a similar size range of vesicles, but with many extremely large aggregates (>5 μm) (data not shown). Sonication of the ultracentrifuged membranes reduced the activity of the Th membranes 4-fold and gave rise to many small vesicles <300 nm in size. Subsequent centrifugation of this sonicated sample restored full activity (Figure 3) and gave rise to vesicles ranging in size between 10 and 1500 nm with some larger aggregates (>2mm). These data suggest that homotypic adhesion may be involved in the formation of Th membrane vesicle aggregates and that the size of the aggregates is directly related to the activity of the Th membranes. These studies demonstrate that extensive crosslinking or simultaneous occupation of multiple ligands on the B cell surface may be necessary for inducing resting B cell proliferation.

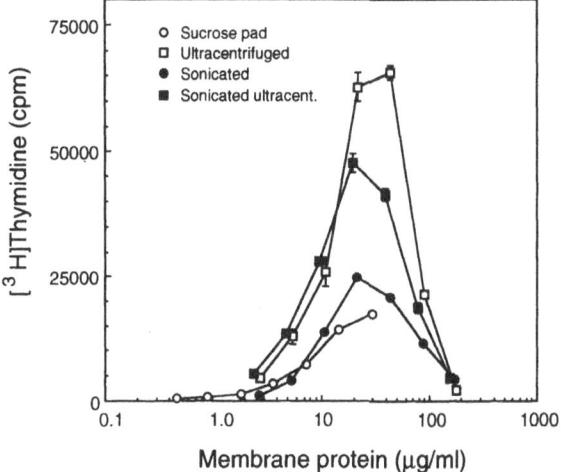

Figure 3. Activity of sonicated and ultracentrifuged Th membranes. Plasma membranes were titrated and incubated for 72 hours with 2×10^4 dense BALB/c B cells. [^{3}H]Thymidine (1 μCi/well) was added for the last 4 hours. Each point is the mean ± standard error of triplicate wells. Samples were freshly prepared Th membrane vesicles from a sucrose pad (open circles); Th membrane vesicles after ultracentrifugation (open squares); vesicles sonicated after ultracentrifugation (closed circles); and sonicated vesicles after ultracentrifugation (closed squares).

DIFFERENTIATION TO ANTIBODY SYNTHESIS

The Requirement for Soluble Lymphokines

Membranes from activated Th cells seemed to deliver the requisite contact signals to resting B cells for inducing several rounds of DNA synthesis.[6] Would the Th membranes also reproduce the function of the intact Th cell in inducing differentiation to secrete immunoglobulin? As shown in Figure 4, resting B cells stimulated with Th membranes produced no immunoglobulin after 7 days of culture unless a lymphokine-containing supernatant from Th2 cells (D10 sn) was provided in the cultures.[6] Th1 lymphokine-containing supernatants were ineffective (Figure 4).[7] Similarly, purified recombinant lymphokines IL-1, IL-2, IL-6, IFN-gamma, or the combination of IL-2 and IFN-gamma did not stimulate immunoglobulin production. These signals were partially provided by IL-4 or IL-5, and the activity in the Th2 (D10) supernatant was completely replaced by a combination of IL-4 and IL-5 (Figure 4).[6,7] Both the amounts and isotypes produced were similar to that induced by B cells stimulated in an antigen-dependent manner by intact Th2 cells from which the lymphokines were derived.[6,12] The major isotypes synthesized in these Th2-dependent B cell responses were IgM, IgG1, and IgE (Figure 4).[6]

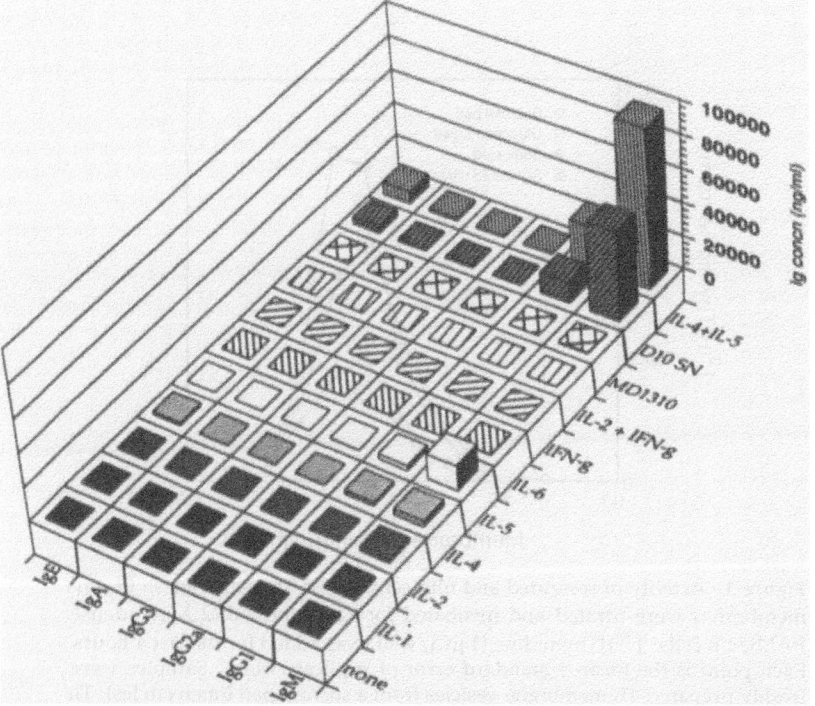

Figure 4. Antibody production by dense B cells stimulated with D10 membranes and various lymphokines.[6] Small dense B cells were cultured for 7 days in the presence of membranes from activated D10 cells and various lymphokines or combinations of lymphokines as shown. Isotype ELISAs for IgM, IgG1, IgG2a, IgG3, IgA, and IgE were as described.[6] Antibody concentrations are the mean of individual ELISA measurements on triplicate cultures and are in nanograms/ml. MD1310, lymphokines from Con A-stimulated MD13-10 Th1 cells; D10 sn, lymphokines from Con A-stimulated D10 Th2 cells. Reprinted by permission from (1).

An additional series of experiments demonstrated that, although Th1 cell clones are poor inducers of immunoglobulin secretion, the contact-dependent signals delivered by Th1 and Th2 cells to B cells are virtually indistinguishable.[7] Thus, it was shown that the lymphokine repertoire of the Th1 cell is solely responsible for its inability to stimulate B cell differentiation.[7,13] Together, the studies in Figure 4[6] and those on Th1 and Th2 membranes[7] showed that the Th cell-derived contact signal stimulates B cell activation and proliferation while Th2-derived lymphokines are required for B cell differentiation to secrete immunoglobulin. It is now possible to completely reproduce the effects of the intact Th cell in inducing B cell responses by using membranes from activated Th cells, IL-4, and IL-5.[6-8]

Differentiation to Antibody Synthesis is B Cell Density Dependent

We have used the Th membrane system to further characterize the induction of B cell differentiation and isotype switching by Th2 lymphokines. One interesting result was that the levels of some isotypes, primarily IgE, were significantly affected by the density of B cells in the cultures (Figure 5). For an input B cell number between 1,000 and 10,000 cells/well, concentrations of IgM, IgG1, and IgE produced by Th membrane-stimulated cultures were directly proportional to input B cell number. In this linear range, similar numbers of B cells switched to secrete each of these isotypes. However, below 1,000 cells/well the levels of each isotype were much less than that predicted for a proportional response (dashed line). Thus, it is possible that signals other than those provided by the Th2 lymphokines, perhaps derived from the interaction of the B cells themselves, may be necessary to induce differentiation to immunoglobulin production.

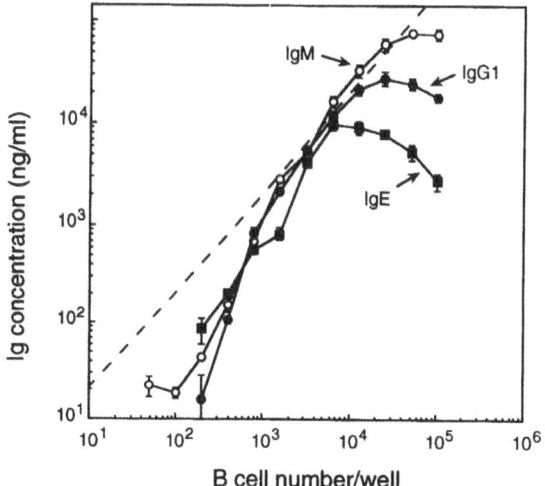

Figure 5. The effect of B cell density on immunoglobulin isotype secretion. Various numbers of B cells were treated with D10 membranes and D10 sn. After 7 days, concentrations of IgM, IgG1, and IgE in the culture medium were measured as described previously.[6] Each point is the mean ± the standard error of individual assays on triplicate cultures.

At high B cell number, IgE levels were preferentially inhibited with much less effect on IgM levels (Figure 5). The effect on IgG1 was intermediate to that of IgM and IgE. One explanation for these results is that at high cell density, medium components are limiting, and IgE may be the first isotype to be affected by cell density because additional cell proliferation may be required for switching to IgE as compared to IgM and IgG1 (see below).[15] Thus, culture conditions can significantly influence the frequency and specificity of isotype switching and *in vitro* B cell responses.

The Window of Lymphokine Sensitivity

Time course experiments were performed to determine when, during Th membrane stimulation of B cells, lymphokines were required for inducing differentiation to antibody secretion. To determine how late lymphokines could be added to Th membrane-stimulated B cells, a Th2 sn was added at various times after initiation of the culture up to day 7. If lymphokine addition was delayed by more than 18 hours, the amounts of IgM and IgE produced were substantially decreased. If the addition of lymphokines was delayed by more than 48 hours, the levels of IgM and IgE were reduced to less than 10% of the maximum levels (Figure 6A).

The converse experiment of removing D10 sn after various times was done by washing cultured B cells and incubated for the remaining culture period with Th membranes in the presence or absence of lymphokines. Removal of lymphokines prior to 72 hours resulted in complete abrogation of IgE secretion and reduction of IgM secretion to 10% of the maximal level. Removal of lymphokines after 72 hours differentially affected the isotypes so that the time of lymphokine removal when 50% of maximal antibody levels were recovered was 84 hours for IgM and 108 hours for IgE (Figure 6B).

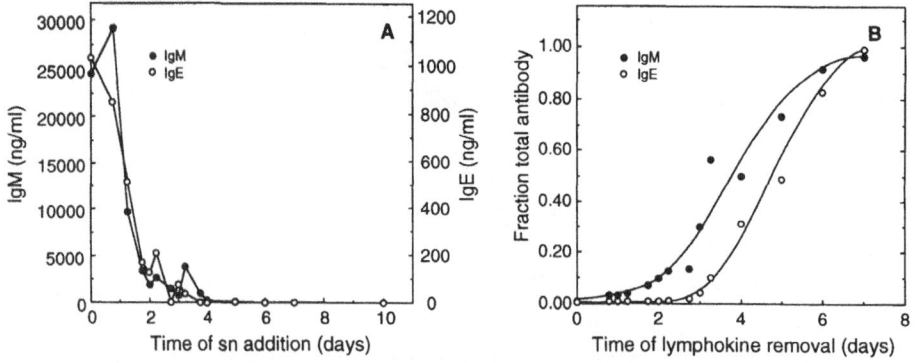

Figure 6. Time course of lymphokine addition and removal from Th membrane-stimulated B cells. **A,** Lymphokine addition time course. Dense B cells were incubated with Th1 membranes, and at the indicated times, D10 sn was added to triplicate cultures. Cultures were harvested after 10 days, and concentrations of IgM (closed circles) and IgE (open circles) determined. **B,** Lymphokine removal time course. Dense B cells were incubated with Th1 membranes and D10 sn. At the indicated times, D10 sn was removed from half of the cells. Cultures were harvested after 7 days and concentrations of IgM (closed circles) and IgE (open circles) determined.[6] For each time point, fraction of antibody produced when D10 sn was removed relative to when D10 sn was present over the entire culture period was calculated.

Thus, the window of time when lymphokines are required to effect differentiation of Th membrane-stimulated B cells is approximately between 1 and

4 days. More importantly, B cells require a longer exposure to lymphokines in order to differentiate and switch to IgE production. This could reflect the fact that, in the presence of IL-4, B cells may be sequentially switching from mu to gamma 1 to epsilon[14,15] and may need to go through additional DNA replication to effect this switching. IL-4 is known to enhance and prolong B cell proliferation,[16] and these results are consistent with the current hypothesis of heavy chain class switching that DNA synthesis is required for the occurrence of lymphokine-dependent recombination.[17-19] The overall picture emerges of a requirement for extended lymphokine exposure to sequentially switch heavy chain class by a mechanism that depends on DNA replication. Not surprisingly, cells producing IgE are rarely generated during normal immune responses.

OVERVIEW AND DISCUSSION

Studies of B cell activation and differentiation using Th membranes have substantially increased our understanding of these processes.[1] It is clear that extended contact between activated Th cells and B cells activates the B cell and drives its entry into the cell cycle. Several programmed rounds of DNA synthesis are initiated by Th cell-B cell interaction, after which the B cell ceases dividing. Productive Th cell contact-dependent signaling also involves crosslinking of a B cell surface ligand that may be CD40 or another unidentified molecule. In the Th membrane system, this B cell surface crosslinking is promoted by homotypic Th cell adhesion induced following Th cell activation and, by itself, is sufficient to induce B cell proliferation. Productive Th cell-B cell contact may also be promoted by adhesion molecules.

Lymphokines secreted by activated Th cells induce differentiation of the B cell to immunoglobulin secretion and induce isotype switching. While not in themselves sufficient to stimulate B cell proliferation, signals delivered by Th2 lymphokines are required shortly after the initial contact-dependent signals, prior to and during DNA synthesis, to induce immunoglobulin secretion. Additionally, there may be a role for activation-induced homotypic B cell adhesion in promoting immunoglobulin synthesis. The extent of lymphokine-dependent signaling may determine the extent to which the B cell switches isotypes and undergoes DNA replication-dependent recombination of the heavy chain locus.

Previously, we have fit this view of B cell activation and differentiation with the existing evidence and the current models of Th cell-dependent B cell responses.[1] The question arose of how the specificity of the B cell response is maintained if the only antigen-specific step is activation of the Th cell (see Figure 1). The studies presented here provide some answers to this paradox. First, we assume a primary role for antigen-specific MHC-restricted Th cell-B cell interaction in the initiation of the B cell response. Second, our studies have demonstrated that after Th cell activation, productive but nonspecific B cell activation results only after extended times of contact with Th cells; the probability that the specific antigen-presenting B cell will be activated is thus highly increased, and the likelihood of a productive casual interaction with a nonspecific B cell is greatly decreased. Importantly, *in vivo* B cell responses do, indeed, contain a nonspecific component that most likely results from this ability of activated bystander Th cells to polyclonally activate B cells.[2]

References

1. P.D. Hodgkin and M.R. Kehry, A new view of B cell activation, *in*: "Advances in Molecular and Cellular Immunology," B. Singh, ed., JAI Press, Inc., Greenwich, Connecticut (1991).
2. B.J.Whalen, H.-P. Tony, and D.C. Parker, Characterization of the effector mechanism of help for antigen-presenting bystander resting B cell growth mediated by IA-restricted Th2 helper T cell lines, *J. Immunol.* 141:2230 (1988).
3. T. Owen, A noncognate interaction with anti-receptor antibody-activated helper T cells induces small resting murine B cells to proliferate and to secrete antibody, *Eur. J. Immunol.* 18:395 (1988).
4. M.H. Julius, H.G. Rammensee, M.J. Ratcliffe, M.C. Lamers, J. Langhorne, and G. Kohler, The molecular interactions with helper T cells which limit antigen-specific B cell differentiation. *Eur. J. Immunol.* 18:381 (1988).
5. R.J. Noelle, J. McCann, L. Marshall, and W.C. Bartlett, Cognate interactions between helper T cells and B cells. III. Contact-dependent, lymphokine-independent induction of B cell cycle entry by activated helper T cells, *J. Immunol.* 140:1807 (1989).
6. P.D. Hodgkin, L.C. Yamashita, R.L. Coffman, and M.R. Kehry, Separation of events mediating B cell proliferation and Ig production by using T cell membranes and lymphokines, *J. Immunol.* 145:2025 (1990).
7. P.D. Hodgkin, L.C. Yamashita, B. Seymour, R.L. Coffman, and M.R. Kehry, Membranes from both Th1 and Th2 T cell clones stimulate B cell proliferation and prepare B cells for lymphokine-induced differentiation to secrete Ig, *J. Immunol.* 147: (1991).
8. R.J. Noelle, J. Daum, W.C. Bartlett, J. McCann, and D.M. Shepherd, Cognate interactions between helper T cells and B cells. V. Reconstitution of T helper function using purified plasma membranes from activated Th1 and Th2 T helper cells and lymphokines, *J. Immunol.* 146:1118 (1991).
9. A.A. Brian, Stimulation of B-cell proliferation by membrane-associated molecules from activated T cells, *Proc. Natl. Acad. Sci. USA* 85:564 (1988).
10. R.J. Noelle, and C.E. Snow, Cognate interaction between helper T cells and B cells, *Immunol. Today* 11:361 (1990).
11. T.B. Barrett, G. Shu, and E.A. Clark, CD40 signaling activates CD11a/CD18(LFA-1)-mediated adhesion in B cells, *J. Immunol.* 146:1722 (1991).
12. T.R. Mossman, and R.L. Coffman, TH1 and TH2 cells: Different patterns of lymphokine secretion lead to different functional properties, *Annu. Rev. Immunol.* 7:125 (1989).
13. R.L. Coffman, B.W.P. Seymour, D.A. Lebman, D.D. Hiraki, J.A. Christiansen, B. Shrader, H.M. Cherwinski, H.F.J. Savelkoul, F.D. Finkelman, M.W. Bond, and T.R. Mosmann, The role of helper T cell products in mouse B cell differentiation and isotype regulation, *Immunol. Rev.* 102:5 (1988).
14. K. Yoshida, M. Matsuoka, S. Usuda, A. Mori, K. Ishizaka, and H. Sakano, Immunoglobulin switch circular DNA in the mouse infected with *Nippostrongylus brasiliensis*: Evidence for successive class switching from μ to ε via γ1, *Proc. Natl. Acad. Sci. USA* 87:7829 (1990).
15. D.A. Lebman, and R.L. Coffman, Interleukin 4 causes isotype switching to IgE in T cell-stimulated clonal B cell cultures, *J. Exp. Med.* 168:853 (1988).
16. P.D. Hodgkin, N.F. Go, J.E. Cupp, and M. Howard, Interleukin-4 enhances anti-IgM stimulation of B cells by improving cell viability and by increasing the sensitivity of B cells to the anti-IgM signal, *Cell. Immunol.* 134:14 (1991).
17. J. Stavnezer, G. Radcliffe, Y.-C. Lin, J. Nietupski, L. Berggren, R. Sitia, and E. Severinson, Immunoglobulin heavy-chain switching may be directed by prior induction of transcripts from constant-region genes, *Proc. Natl. Acad. Sci. USA* 85:7704 (1988).
18. W. Dunnick, M. Wilson, and J. Stavnezer, Mutations, duplication, and deletion of recombined switch regions suggest a role for DNA replication in the immunoglobulin heavy-chain switch, *Molec. Cell. Biol.* 9:1850 (1989).
19. W. Dunnick, and J. Stavnezer, Copy choice mechanism of immunoglobulin heavy-chain switch recombination, *Molec. Cell. Biol.* 10:397 (1990).

THE LOW AFFINITY IgE Fc RECEPTOR (CD23) PARTICIPATES IN B CELL ACTIVATION

Thomas J. Waldschmidt and Lorraine T. Tygrett

Department of Pathology
University of Iowa
College of Medicine
Iowa City, Iowa 52242

INTRODUCTION

The low affinity IgE Fc receptor (FcεRII or CD23) is expressed on all mature B lymphocytes (reviewed in 1 and 2). The FcεRII is a 45 to 49 KD glycoprotein, and is categorized as a class II transmembrane molecule on account of its inverted membrane orientation. In addition to the transmembrane form of the receptor, a soluble form is readily released from B cells, a result of cleavage by a cell surface protease. cDNA cloning has revealed that the FcεRII is a member of the hepatic lectin family, a somewhat unusual finding since all other characterized Fc receptors are known to be members of the immunoglobulin (Ig) gene superfamily.

To date, no clear function for the surface form of the B cell FcεRII has been identified. Most investigators have focused on the role of the cleaved receptor, and have shown that the soluble FcεRII can serve as a growth co-factor for immature and mature T cells, myeloid precursors, and B cells, prevent macrophage migration, and potentiate IgE secretion (reviewed in 2). Although no dominant role for the surface FcεRII has been found, several groups have demonstrated that crosslinking this receptor with IgE or anti-CD23 antibodies downregulates IgE secretion (3,4), while other investigators have reported that IgE immune complexes can be effectively taken up by B cells resulting in processing and presentation of the complexed antigen (5,6).

Our laboratory has taken another approach in searching for the function of the FcεRII. This approach is based upon the known function of the other Fc receptor expressed on B cells, the IgG Fc receptor (FcγRII or CD32). A large body of evidence has demonstrated that the FcγRII delivers a strong inhibitory signal to the B cell when crosslinked with surface Ig, and can effect downregulation of B cell activation, proliferation, and differentiation (7). The inhibition only occurs if the FcγRII is crosslinked to surface Ig, since crosslinking the FcγRII with itself has no demonstrable effect on B cells. This sIg-FcγRII crosslinking can be performed experimentally with intact rabbit IgG anti-mouse IgM or IgD, where the Fc portion of the rabbit antibody binds to the FcγRII and the antigen binding sites of the molecule bind to surface Ig. In vivo, this crosslinking is thought to be mediated by IgG immune complexes or anti-idiotypic antibodies, and has been proposed as a means of antibody feedback regulation.

Based upon the ability of the FcγRII to exert its function when crosslinked with surface Ig, we sought to examine whether the FcεRII might also deliver a negative signal to the B cell when similarly crosslinked. Accordingly, we developed a unique reagent which would allow one to test the effect of sIg-FcεRII crosslinking, and thereby lend information as to the function of this receptor on B cells.

Mechanisms of Lymphocyte Activation and Immune Regulation IV: Cellular Communications
Edited by S. Gupta and T.A. Waldmann, Plenum Press, New York, 1992

DEVELOPMENT OF A MOUSE IgE ANTI-MOUSE IgD

The 10.4.22 hybridoma cell line secretes a mouse IgG2a anti-mouse IgD[a] antibody (8). This anti-allotypic hybridoma was developed some years ago by Herzenberg and colleagues, and was chosen for our study since it secretes an antibody of the IgG2a isotype, is reactive with surface Ig, and in soluble form, is not mitogenic for B cells. It is well known that all hybridoma cell lines produce isotype switch variants at a low frequency (9). In the case of the IgG2a producing 10.4.22 cell line, IgE switch variants would be generated owing to the location of the epsilon heavy chain gene directly downstream from the γ2a locus. We therefore attempted to isolate such a variant, since the resulting IgE anti-IgD antibody would have the potential to crosslink the FcεRII with surface Ig. An IgE switch variant of the 10.4.22 cell line was isolated by sorting cells which were negative for surface IgG into microtiter wells, allowing the cells to form clones, and identifying any wells containing IgE antibody with an IgE-specific ELISA. Figure 1 illustrates the staining pattern of the parent 10.4.22 cell line, and shows the sorting gate from which the surface IgG negative hybridoma cells were selected.

In addition to its reactivity with the IgE specific ELISA, the antibody secreted by the 10.4.22 switch variant was shown to be composed of epsilon heavy chains by virtue of its slower migration on an SDS-polyacrylamide gel. (The epsilon heavy chain has 4 constant region domains, and is therefore easily distinguishable from gamma heavy chains). Using flow cytometry, the IgE variant protein was also shown to retain its IgD[a] binding activity, and more important, was shown to bind to the FcεRII. Thus, the 10.4.22 switch variant secretes a mouse IgE anti-mouse IgD[a] antibody, a reagent which should now allow one to test the functional consequences of crosslinking the FcεRII with surface Ig.

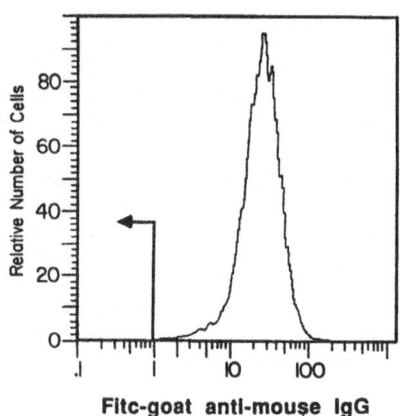

Figure 1. Sort cloning of IgG-10.4.22 cells. 10.4.22 hybridoma cells were stained with a goat anti-mouse IgG. Cells failing to stain with the antibody (0.04%) were sorted into microtiter wells at 2 cells per well.

MITOGENIC ACTIVITY OF THE 10.4.22 IgE SWITCH VARIANT

In an initial series of experiments, the 10.4.22 IgG parent and IgE variant proteins were tested for their ability to stimulate B cells. The IgG protein was purified on a protein A-Sepharose column, and the IgE antibody was purified on a monoclonal rat anti-mouse IgE (EM95)-Sepharose column. All preparations of antibody were shown to contain less than 7.5 ng/ml of endotoxin by the Limulus assay. Resting B cells were purified from either DBA/2 (Igh[a]) or C57Bl/6 (Igh[b]) mice. Small B cells were prepared by T cell depletion followed by isolation on a discontinuous Percoll gradient (70-75% interface). When incubating the Igh[a] B cells with increasing concentrations of the parent IgG anti-IgD antibody, one observes no stimulation as measured by thymidine incorporation (Table I).

This is consistent with previous reports demonstrating that the 10.4.22 antibody is not mitogenic in soluble form (10). When incubating the cells with the IgE variant protein however, there is a striking proliferative response. This is in stark contrast to the IgG parent antibody, and immediately indicates that crosslinking the FcεRII with surface Ig does not mimic the inhibitory effect of FcγRII-Ig crosslinking.

TABLE I

Effect of the 10.4.22 IgG and IgE Antibodies on DBA/2 Igh[a] B Cells [a]

Stimulus	Conc. (ug/ml)	CPM ± S.D.
Media alone	-	1900 ± 468
ε variant	5.0	38,967 ± 290
ε variant	10.0	35,646 ± 2973
ε variant	25.0	75,350 ± 3649
ε variant	50.0	91,916 ± 4586
ε variant	100	124,245 ± 6718
γ parent	5.0	2111 ± 227
γ parent	10.0	2800 ± 361
γ parent	25.0	3133 ± 368
γ parent	50.0	3859 ± 364
γ parent	100	3977 ± 281

[a] B cells were prepared by treating whole spleen cells with anti-Thy 1.2 (HO13.4) plus complement, followed by centrifugation through a discontinuous Percoll gradient. Cells settling at the 70-75% Percoll interface were harvested and incubated with the various stimuli for 48 hours at 1×10^5 cells per well in 96 well plates. Cells were pulsed with 0.5 μCi of ^{3}H-Thymidine during the last 4 hours of culture.

Since the 10.4.22 antibodies are specific for B cells of the Igh[a] haplotype, a convenient specificity control is to test the effect of the parent and switch variant proteins on B cells of the Igh[b] haplotype. This experiment is shown in Table II, and demonstrates that there is no significant activity of either the IgG or IgE anti-IgD antibodies. This latter experiment shows that the stimulatory effect of the IgE switch variant is not simply due to engaging the FcεRII, but also requires the binding of surface IgD.

TABLE II

Effect of the 10.4.22 IgG and IgE Antibodies on C57Bl/6 Igh[b] B Cells [a]

Stimulus	Conc. (ug/ml)	CPM ± S.D.
Media alone	-	997 ± 196
ε variant	5.0	1828 ± 357
ε variant	10.0	3143 ± 491
ε variant	25.0	3554 ± 512
ε variant	50.0	5263 ± 435
γ parent	5.0	947 ± 152
γ parent	10.0	1485 ± 548
γ parent	25.0	1068 ± 82
γ parent	50.0	997 ± 145

[a] B cells were prepared and incubated as described in the legend to Table I.

CELL CYCLE ANALYSIS OF IgE ANTI-IgD STIMULATED B CELLS

The experiment represented in Table I examined the ability of B cells to proliferate in response to the 10.4.22 antibodies. In order to more carefully examine the response to the IgG and IgE anti-IgD proteins, we performed Acridine Orange (AO) analysis on cells stimulated for either 24 or 48 hours. AO staining allows for simultaneous detection of both DNA and RNA, and therefore permits one to precisely determine the percentage of cells in the G0, G1, and S/G2 phases of the cell cycle. LPS stimulated cells were used as a positive control. The results of the experiment are shown in Table III, and are consistent with the data obtained with thymidine incorporation. As expected, cells incubated in media alone remained in G0. B cells stimulated with the 10.4.22 parent antibody did show a small response by 48 hours, where a proportion of the cells did proceed into cycle. When examining the response to the IgE switch variant however, it is clear that at 24 hours most of the cells have left G0, and by 48 hours, many have progressed into the S and G2 phases. It is also noteworthy that the IgE anti-IgD induces virtually all of the cells to enter cell cycle, whereas in the LPS cultures, a significant proportion remain in G0. It has previously been reported that both LPS and conventional anti-Ig reagents have the capacity to induce only a portion of normal resting B cells to enter cell cycle (11). It appears therefore, that the IgE switch variant has exceptional stimulatory properties, a characteristic which may owe itself to its ability to engage the FcεRII.

TABLE III

Cell Cycle Analysis of IgG and IgE anti-IgD Stimulated B Cells [a]

		PERCENTAGE OF CELLS :		
Stimulus	Hours	G0	G1	S/G2
Media	24	98.1	1.0	0
γ parent	24	92.8	5.4	0
ε variant	24	2.6	95.3	1.3
LPS	24	48.1	47.9	1.7
Media	48	97.1	1.0	0
γ parent	48	42.1	44.5	10.5
ε variant	48	3.1	57.3	36.2
LPS	48	21.2	51.5	25.0

[a] B cells were prepared and incubated as described in the legend to Table I. The 10.4.22 IgG and IgE proteins were added at 25 µg/ml. After either 24 or 48 hours of culture, cells were fixed and stained with AO. Cells unaccounted for represent doublets or apoptotic cells.

CAN THE FcγRII ACCOUNT FOR THE DIFFERENTIAL FUNCTIONAL ACTIVITY OF THE PARENT AND VARIANT PROTEINS?

One could argue that the inability of the 10.4.22 IgG parent antibody to induce B cell proliferation is due to engagement of the FcγRII. As discussed in the Introduction, the FcγRII is known to exert a strong inhibitory signal to the B cell when crosslinked with surface Ig. Thus, it is important to ask whether the IgE variant is mitogenic simply because the FcγRII can no longer be engaged. In order to test this question, purified B cells were incubated with saturating concentrations of the anti-FcγRII antibody 2.4G2. This antibody has the capacity to block ligand binding by the FcγRII, and should disallow the IgG anti-IgD parent protein from binding this receptor. Accordingly, if the stimulatory capacity of the IgE variant antibody is due to the lack of FcγRII involvement, then B cells cultured with the IgG parent plus 2.4G2 should now respond in a positive fashion. In data not shown, the anti-FcγRII antibody did not reverse the non-stimulatory nature of the 10.4.22 parent protein. This indicates that the stimulatory capacity of the IgE anti-IgD antibody is not due to the absence of negative signalling by the FcγRII, but may indeed be due to the positive participation of the FcεRII.

CO-CROSSLINKING OF THE SURFACE Ig AND FcεRII IS REQUIRED FOR B CELL ACTIVATION

Although the 10.4.22 IgE variant protein have the ability to engage both surface IgD and the FcεRII, the question arises as to whether the B cell activation is due to co-crosslinking of these two surface molecules, or to simultaneous yet independent crosslinking of surface Ig and the FcεRII. If the latter situation is true, then one should be able to reproduce the effect of the IgE anti-IgD by simultaneously culturing B cells with the 10.4.22 IgG parent antibody and with an irrelevant IgE protein. The results of such an experiment are shown in Table IV, in which several permutations were tested. B cells were incubated with biotin-conjugated IgE (A3B1, an IgE anti-TNP hybridoma protein) and avidin, the IgG anti-IgD (in the presence of 2.4G2 to eliminate any effects of the FcγRII), and in some instances, optimal levels of Interleukin-4 (IL-4). IL-4 was included since it upregulates the expression of the FcεRII, and is also a growth co-factor for anti-Ig stimulated B cells. With this combination of reagents, in which surface IgD is crosslinked with the 10.4.22 IgG parent protein, and the FcεRII is crosslinked with the biotin-IgE plus avidin, no significant stimulation is observed. Even in the presence of IL-4, no response is observed. Thus, the strong activation induced by the IgE anti-IgD antibody is apparently due to the co-crosslinking of the sIg and FcεRII, and not due to synergistic signals generated by independent crosslinking of these two molecules.

TABLE IV

The Effect of Independent Crosslinking of Surface IgD and FcεRII on B Cells [a]

Biotin-IgE (μg/ml)	10.4.22 γ parent (μg/ml)	Avidin (μg/ml)	rIL-4 50 U/ml	CPM ± S.D.
-	-	-	-	484 ± 26
-	-	-	+	1183 ± 19
10.0	10.0	5.0	-	253 ± 122
50.0	10.0	5.0	-	401 ± 57
-	10.0	-	+	3651 ± 445
10.0	10.0	5.0	+	1644 ± 335
50.0	10.0	5.0	+	3662 ± 542

[a] B cells were prepared and incubated as described in the legend to Table I. All wells contained saturating (200 μg/ml) levels of the monoclonal anti-FcγRII antibody 2.4G2.

BLOCKING THE FcεRII DIMINISHES THE STIMULATORY CAPACITY OF THE 10.4.22 IgE VARIANT

The data presented thus far has strongly indicated that engagement of the FcεRII is essential to the stimulatory capacity of the 10.4.22 IgE variant antibody. If this is indeed the case, then blocking the ability of the IgE anti-IgD protein to bind to the FcεRII should greatly reduce its ability to activate B cells. We therefore tested whether the monoclonal anti-FcεRII antibody B3B4 (12) could reduce the activity of the IgE anti-IgD. B3B4 binds at or near the ligand binding site, and has the capacity to block the binding of soluble IgE to the FcεRII. When added in significant excess however, B3B4 was unable to alter the proliferative activity of the 10.4.22 IgE variant antibody (data not shown).

Although this result was unexpected, one must consider the potential binding characteristics of the 10.4.22 IgE protein. Once bound to the surface of the B cell, a single IgE anti-IgD antibody has the potential to bind two FcεRII molecules as well as two IgD molecules. [IgE has the capacity to bind two FcεRII on account of a binding site displayed on each of the epsilon heavy chains (13)]. Thus, the IgE anti-IgD is likely to interact with the surface B cells via four point binding, and display an overall avidity for the B cell which is too high for the B3B4 antibody to compete with.

In order to effectively compete with the 10.4.22 IgE variant, we developed a reagent which not only binds to the FcεRII, but also has the capacity to bind to the surface of B cells using four point binding. Specifically, we isolated an IgE switch variant from the mouse IgG2a anti-K^dD^d (MHC class I) hybridoma cell line 34-1-2S. This IgE anti-class I variant was isolated using the same sort cloning strategy as described for production of the 10.4.22 variant. In addition to its ability to engage the FcεRII, this 34-1-2S IgE variant binds surface class I molecules and should likewise interact with the surface of B cells with a high avidity. As shown in Table V, excess concentrations of the IgE anti-class I reagent caused a significant reduction in the proliferative response of IgE anti-IgD stimulated B cells (up to 59% inhibition). Thus as predicted, disruption of sIg-FcεRII crosslinking results in a reduction of the stimulatory capacity of the 10.4.22 IgE. This result reinforces the notion that the FcεRII may indeed play a role in B cell activation.

TABLE V

The Ability of the IgE Anti-class I Protein to Inhibit the Stimulatory Capacity of the 10.4.22 ε Variant [a]

10.4.22 ε variant (μg/ml)	34-1-2S ε variant (μg/ml)	rIL-4 50 U/ml	CPM ± S.D.
-	-	-	1869 ± 789
-	-	+	5234 ± 1409
0.1	-	-	33,477 ± 5504
0.1	-	+	54,071 ± 5501
1.0	-	-	100,177 ± 4731
1.0	-	+	112,092 ± 4828
0.1	25.0	-	32,155 ± 3617
0.1	25.0	+	30,001 ± 2572
1.0	25.0	-	53,243 ± 4931
1.0	25.0	+	46,478 ± 4917

[a] B cells were prepared and incubated as described in the legend to Table I.

DISCUSSION

The aim of this study was to test whether the FcεRII functions in a manner similar to the FcγRII. It is well established that the FcγRII mediates a potent negative signal to B cells when crosslinked to sIg. In order to ascertain whether the FcεRII also provides regulatory signals when crosslinked to sIg, we developed a switch variant of the 10.4.22 cell line resulting in an IgE anti-IgD antibody. When testing this reagent on purified resting B cells, it did not function to downregulate B cell activity, but rather, invoked a strong positive response. The IgE anti-IgD was shown to trigger a rapid entry into cell cycle and subsequent progression through the S and G2 phases. The stimulatory capacity of the 10.4.22 variant was not due to the absence of FcγRII mediated inhibition, an event potentially invoked by the parent IgG anti-IgD antibody, since blocking this receptor did not reverse the non-stimulatory nature of the parent protein. Additional experiments demonstrated that the FcεRII had to be directly crosslinked with sIg, since independent crosslinking of the two surface molecules failed to generate the same proliferative response. Finally, the stimulatory capacity of the IgE anti-IgD was greatly diminished by excess concentrations of an IgE anti-class I antibody, strongly implicating the participation of the FcεRII in the activation events generated by the 10.4.22 IgE variant.

The results of this study clearly demonstrate that the B cell FcεRII and FcγRII do not share the same function. Whereas the FcγRII serves to downregulate B cell activity, the present data suggest that the FcεRII may upregulate the response of these cells. Clearly, further work is required to ascertain the precise role of the FcεRII in the activation cascades invoked by sIg-FcεRII crosslinking. It is certainly of interest to explore what additional signal transduction mechanisms may be employed by the B cell when the FcεRII is a part of the activation complex. To this end, Yodoi and co-workers have recently reported that the tyrosine kinase p59^{fyn} was physically associated with the FcεRII in human cells (14).

A more important question however, is the physiologic significance of these findings. The IgE anti-IgD system was designed to serve as a model of IgE immune complex regulation, since sIg-FcεRII crosslinking is mediated by these complexes in vivo. If IgE immune complexes do indeed serve to activate or enhance the response of antigen specific B cells, then one must further question whether we understand all the roles that IgE might play in the humoral response. In addition to arming mast cells to protect skin and mucosal surfaces against parasitic invasion, IgE, in the form of immune complexes, might also serve as an adjuvant early in the immune response.

ACKNOWLEDGMENTS

The authors wish to thank Ms. Teresa Duling and Mr. Patrick Jacobs for their expert operation of the flow cytometer. This work is supported by research grants NIH R29AI31265 and Council for Tobacco Research Grant 2785.

REFERENCES.

1. D.H. Conrad, FcεRII/CD23: The Low Affinity Receptor for IgE, Ann. Rev. Immunol. 8:623 (1990).
2. G. Delespesse, U. Suter, D. Mossalayi, B. Bettler, M. Sarfati, H. Hofstetter, E. Kilcherr, P. Debre, and A. Dalloul, Expression, Structure, and Function of the CD23 Antigen, Adv. Immunol. 49:149 (1991).
3. E. Sherr, E. Macy, H. Kimata, M. Gilly, and A. Saxon, Binding the Low Affinity FcεR on B Cells Suppresses Ongoing Human IgE Synthesis, J. Immunol. 142:481 (1989).
4. H. Luo, H. Hofstetter, J. Banchereau, and G. Delesspesse, Cross-linking of CD23 Antigen by its Natural Ligand (IgE) or by Anti-CD23 Antibody Prevents B Lymphocyte Proliferation and Differentiation, J. Immunol. 146:2122 (1991).
5. M.R. Kehry and L.C. Yamashita, Low-affinity IgE Receptor (CD23) Function on Mouse B Cells: Role in IgE-dependent Antigen Focusing, Proc. Natl. Acad. Sci. U.S.A. 86:7556 (1989).
6. U. Pirron, T. Schlunch, J.C. Prinz, and E.P. Rieber, IgE-dependent Antigen Focusing by Human B Lymphocytes is Mediated by the Low-affinity Receptor for IgE, Eur. J. Immunol. 20:1547 (1990).
7. N.R. StC. Sinclair, Immunoregulation by Antibody and Antigen-antibody Complexes, Trans. Proc. 10:2 (1978).
8. V.T. Oi, P.P. Jones, J.W. Goding, L.A. Herzenberg, and L.A. Herzenberg, Properties of Monoclonal Antibodies to Mouse Ig Allotypes, H-2, and Ia Antigens, Curr. Top. Microbiol. Immunol. 81:115 (1978).
9. J. L. Dangl and L.A. Herzenberg, Selection of Hybridomas and Hybridoma Variants Using the Fluorescence Activated Cell Sorter, J. Immunol. Meth. 52:1 (1982).
10. D.K. Goroff, A.L. Stall, J.J. Mond, and F.D. Finkelman, In Vitro and In Vivo B Lymphocyte-Activating Properties of Monoclonal Anti-δ Antibodies I. Determinants of B Lymphocyte-Activating Properties, J. Immunol. 136:2382 (1986).
11. H. Seyschab, R. Friedl, D. Schindler, H. Hoehn, P.S. Rabinovitch, and U. Chen, The Effects of Bacterial Lipopolysaccharide, Anti-receptor Antibodies and Recombinant Interferon on Mouse B Cell Cycle Progression Using 5-bromo-2'-deoxyuridine/Hoechst 33258 Dye Flow Cytometry, Eur. J. Immunol. 19: 1605 (1989).

12. M. Rao, W.T. Lee, and D. H. Conrad, Characterization of a Monoclonal Antibody Directed Against the Murine B Lymphocyte Receptor for IgE, J. Immunol. 138:1845 (1987).
13. W.T. Lee and D.H. Conrad, The Murine Lymphocyte Receptor for IgE II. Characterization of the Multivalent Nature of the B Lymphocyte Receptor for IgE, J. Exp. Med. 159:1790 (1984).
14. K. Sugie, T. Kawakami, Y. Maeda, T. Kawabe, A. Uchida, and J. Yodoi, Fyn Tyrosine Kinase Associated with FcεRII/CD23: Possible Multiple Roles in Lymphocyte Activation, Proc. Natl. Acad. Sci. U.S.A. 88:9132 (1991).

T CELL ADHESION CASCADES:

GENERAL CONSIDERATIONS AND ILLUSTRATION WITH CD31

Yoshiya Tanaka and Stephen Shaw

Experimental Immunology Branch
National Cancer Institute
National Institutes of Health
Bldg 10 Room 4B17
Bethesda, MD 20892

INTRODUCTION

Widespread appreciation of the importance of T cell adhesion has emerged only in the last 5-10 years. But now the topic is being widely discussed and many aspects carefully reviewed.[1-8] This review briefly outlines one important emerging concept, namely that adhesion is not a simple unitary event but rather a co-ordinated sequence of events. This concept is emerging concurrently from studies in a variety of model systems.[3,9-12] We illustrate how this concept has emerged from ours and others studies of T cells. Many aspects of T cell adhesion and biology are touched on which are beyond the scope of this brief presentation; the citations therefore emphasize reviews as a starting point for further reading.

Adhesion is critical to T cell function, both: 1) where it goes and 2) what it does. T cell migration/recirculation through tissue is necessarily dependent on complex sequences of events which involve adhesion: initial interaction with and adhesion to endothelium; migration between endothelial cells, through basement membrane and into the tissue; chemotaxis, haptotaxis and random migration within tissue; and exodus into lymph. The effector function of T cells is likewise dependent on adhesion: effective interactions of T cells during immunological surveillance; the recognition events during priming by an APC; and the strong interaction of effector T cells with their partner cells (targets for CTL or B cells for helper T cell).

T cells need to be able to interact with cells in virtually every tissue and microenvironment in the body. There are two corollaries to this observation. First, T cells must (and do) move incessantly. It is said that on the average T cells move daily from blood to non-lymphoid tissue or to lymphoid tissue. Second, T cells must be endowed with a complement of adhesion molecules which enable them to interact effectively in an extraordinarily diverse range of microenvironments. This fits with the facts that resting T cells express at least 20 adhesion molecules and, when activated, express an even greater range of adhesion molecules.

PARADIGMS OF T CELL ADHESION CASCADE

For the sake of discussion, we identify two functionally different kinds of T cell interactions and use those two interactions to discuss the concept of a T cell adhesion cascade. The first is T cell interaction with endothelium. This is a fundamentally important interaction for the movement of a T cell from blood into the site where it encounters antigen and mediates its effector functions. This is an *antigen-independent* adhesive process. The second is T cell recognition of antigen on the surface of an antigen-positive cell, resulting in *antigen-specific* T cell activation.

Mechanisms of Lymphocyte Activation and Immune Regulation IV: Cellular Communications
Edited by S. Gupta and T.A. Waldmann, Plenum Press, New York, 1992

GLUE

In both the antigen-specific and the antigen-independent T cell adhesion cascades, strong adhesion is a fundamental component, and that strong adhesion is mediated largely by molecules which are members of the integrin family. The integrins are a very diverse and widely distributed family with characteristic dimeric structure; they are of cardinal importance in mediating cell adhesion to other cells and to extracellular matrix components. Each integrin is composed of unique pair of α and β chains. Resting T cells express a minimum of four integrins: LFA-1 (αL/β2), VLA-4 (α4/β1), VLA-5 (α5/β1), VLA-6 (α6/β1) (listed in descending order of median abundance on resting CD4+ circulating T cells).

In these comments we concern ourselves primarily with the integrins LFA-1 and VLA-4. These two integrins, like other integrins, have multiple ligands. LFA-1 binds to: 1) ICAM-1 which is expressed at low levels by resting antigen-presenting cells but at high levels by many cells (including vascular endothelium) exposed to inflammatory cytokines;[13] 2) ICAM-2 which is expressed on many cells including endothelial cells and 3) ICAM-3 which is expressed on many cells including lymphocytes.[14]

LFA-1 is normally involved in both the T/endothelial and the T/APC interactions. For a variety of reasons LFA-1 has dominated much of our thinking about T cell adhesion per se and about adhesion in T cell recognition: it is the most highly expressed T cell integrin; it was the first discovered and well characterized; its effects on in vitro immunologic assays of T cell function are profound. LFA-1 interactions contribute to T cell binding to endothelial cells via binding to both ICAM-1 and ICAM-2.[15,16] However, LFA-1 is NOT indispensable in T cell function; this conclusion is unequivocally demonstrated by the findings in leukocyte adhesion deficiency (LAD);[17] such patients lacking LFA-1 (and other molecules sharing the integrin β2 chain) have profound problems with granulocyte function, but their T cell function and migration is not a serious problem clinically.

VLA-4 binds both to: 1) the relatively ubiquitous extracellular matrix component fibronectin; and 2) the cell surface ligand VCAM-1 expressed principally on inflamed endothelium; 3) an as yet undefined ligand on leukocytes.[1,9,18,19] Its physiological functions undoubtedly include T cell interaction with endothelial cells: both in normal homing and in inflammatory responses.[9,16, 20-22] In addition, VLA-4 can mediate interactions between leukocytes.[18,19] It can potentially contribute to antigen-specific interactions by virtue of its expression on antigen presenting cells in tissue;[23] this possibility is highlighted by the findings that VLA-4/VCAM-1, like LFA-1/ICAM-1 interactions, are profoundly costimulatory with CD3 mAb in activating T cells.[24]

TRIGGER

The expression of integrins at the surface of a T cell is NOT sufficient for them to mediate binding to their ligands. Many circulating T cells express high levels of LFA-1 and VLA-4 but bind poorly to ICAM-1 and VCAM-1. Before T cells can bind efficiently to these ligands, they must be "triggered". There is a growing class of cell surface molecules which have the capacity to trigger functional activation of the integrin molecules without changing their level of expression. The first described was CD3; highly multivalent cross-linking of CD3 on resting T cells induces rapid transient activation of the function of T cell LFA-1.[25] Thus, CD3 is a "trigger" molecule which regulates the function of the integrin glue. Thus, there is a partnership between trigger molecules such as CD3 and integrin molecules like LFA-1 in antigen specific recognition.

Subsequently, this finding has been broadened in numerous dimensions. First, the CD3 trigger induces not only the function of LFA-1, but also of VLA-4 (and other integrins).[26] Second, as predicted from these models, antigen-specific interactions induce augmented integrin function.[27] Third, there are other trigger molecules on T cells, including CD7, CD28, CD44 and CD31.[28-30]

What is the conceptual significance of having multiple adhesion-triggering molecules? There must be specialization of triggering molecules and pathways. Obviously CD3/TCR is not relevant to the interactions of circulating T cells during their antigen-nonspecific interactions with endothelium; much of the adhesion which accompanies their migration through tissues must likewise be regulated by triggering molecules distinct from CD3. We propose that CD31, described below, figures in such antigen-independent regulation of adhesion.[30] In addition, there are likely molecules which collaborate with CD3 in regulating the adhesion in antigen-specific interactions; CD7 and CD28 are likely to be involved in such a way.[28]

Our most recent studies have emphasized CD31.[30] CD31 mAb induce certain resting T cells to bind to integrin ligands. That places CD31 in the category of "trigger" molecules, which are as important in adhesion as are the integrin "glue" molecules. Three features make CD31 an especially interesting trigger molecule. The first is that CD31 has a tissue distribution which is striking in several respects: a) CD31 is expressed by only a subset of T cells. Most, but not all CD8 cells express CD31 at intermediate levels; The few CD8 cells which do NOT express it are among the CD8+CD45RO+ memory cells. Only a minority of CD4 cells express it; those which do are CD4+CD45RA+ naive cells. Thus, CD31 is biased towards expression of CD8 cells and naive cells, but does not conform to the distribution of any other known markers on these T cells. Regulation of expression of a trigger molecule, like CD31, is as important as regulation of expression of an adhesion molecule. Thus, the unique CD31+ subsets of T cells will be endowed with the capacity to adhere in response to encounter with the CD31 ligand (which remains to be conclusively defined), provided they also encounter appropriate integrin ligands. b) CD31 is expressed by essentially all formed elements in the blood apart from the CD31-negative T cell subsets mentioned above. Although it remains to be tested, we postulate that CD31 may also be an adhesion-inducing trigger on a wide variety of cell types. c) CD31 is also expressed on vascular endothelium. It has been inferred to be involved in monocyte interaction with endothelium.[31]

The second unique feature of CD31 is the lack of requirement for extensive crosslinking in the induction of integrin function. All the other trigger molecules on resting T cells can be activated only by mAb binding followed by extensive cross-linking provided by a sandwich reagent like goat anti-mouse IgG. In contrast, CD31 mAb trigger and do so without any sandwich reagent. This suggests that dimerization of CD31 is sufficient to trigger integrin function. Accordingly, CD31 may be a uniquely sensitive trigger molecule.

The third remarkable feature regarding CD31 is the selectivity observed in its induction of integrin function. Previously, there has been no evidence that the different triggers on T cells have any selectivity with regard to which integrins they activate functionally. CD3 and CD31 both induce functional activation of LFA-1 binding to ICAM-1 and VLA-4 binding to VCAM-1. However, quantitative comparisons demonstrate that CD31 preferentially induces the function of VLA-4 while CD3 preferentially induces LFA-1. This has several important implications: there must be partially non-overlapping biochemical mechanisms transducing the CD3 and CD31 signals to provide the selectivity; the potential for sophisticated regulation of adhesion is dramatically increased by the capacity of different trigger molecules to preferentially activate different adhesion effector molecules.

What is the biological function of CD31 on the T cells which express it? We hypothesize that it may be particularly important in interaction of CD31+ T cells with endothelium. The preferential induction of VLA-4 function by the CD31 trigger coincides with the importance of the VLA-4 molecule in T cell binding to endothelium.[9,16,20-22] The sensitivity of the CD31 transduction mechanism is consistent with the need for very rapid induction during T cell encounter with endothelium. Thus, this hypothesis fits the facts that CD31 expression marks most formed elements (with the exception of T cell subsets noted above) in the blood and that CD31 is involved in monocyte binding to endothelium. We are currently testing this hypothesis of CD31 involvement in T/endothelial cell interaction.

Finally, the preferential expression of CD31 on CD8 cells more than CD4 cells together with selective capacity to induce VLA-4 function, prompt us to propose that CD31 may contribute to the preferential movement of CD8 cells into particular tissues and or inflammatory sites. One intriguing possible site for CD31 involvement is gut, where there is a predominance of CD8 cells.[32] In addition to VLA-4 ($\alpha 4\beta 1$), there is another $\alpha 4$-integrin ($\alpha 4\beta 7$) on gut homing T cells which interacts with a ligand on Peyer's patch high endothelial venule (HEV).[33,34] Also the preferential expression of CD31 on naive cells raises the possibility that it may contribute to the process of migration of naive cells into lymph node.[35] Furthermore, given data that CD31 may participate in homophilic interactions with CD31,[36] T cell CD31 may contribute to T cell interaction with CD31+ APC.

TETHER

The foregoing two classes of molecules are fundamental elements in an adhesion cascade:

a strong adhesion molecule and a trigger molecule which regulates that adhesion. However, often the demands of adhesion require more sophistication even than that, and have given rise to sequential utilization of additional adhesive interactions. Cell binding to endothelium requires a molecular interaction between T cell and endothelium even before the trigger molecules get involved. In our view, this provides the first transient contact between cell membranes which enables trigger molecules on the circulating cell to interact with ligands which may be present on the opposing endothelial surface. For interactions with endothelial cells, the unique family of selectin molecules serves this role for virtually all formed elements in the blood, perhaps due to special kinetics of the lectin interaction with carbohydrate.[37] The molecule L-selectin (previously Leu-8, Lam-1, Mel-14), expressed on most T cells, is involved in T cell binding to endothelium; this requirement is observed primarily when the assay is performed with the T cells in motion,[38] thus mimicking at least crudely the situation of flow present in vivo. Although originally proposed as a homing receptor for T cell movement into lymph node,[39] L-selectin seems to be important more generally in T cell interaction with endothelium.[40] Of note, the engagement of selectin molecules on T cells do not trigger T cell adhesion to endothelium.

In T cell interaction with APC (or other collaborating cell in antigen-specific recognition), we believe that there are also antigen-independent adhesion processes which precede engagement of the T cell receptor;[41] these serve the analogous role of enabling a period of contact between T cell and APC during which the T cell receptor will encounter MHC/peptide molecules on the opposing cell.

CONCLUSIONS

Thus T cell adhesion events are co-ordinated sequences of adhesive interactions. For T cell binding to endothelium, a prototypic sequence of events for a successful antigen-nonspecific interaction would be: 1) L-selectin binding to a specific endothelial carbohydrate, resulting in transient tethering of the T cell to the endothelium; 2) Encounter between trigger molecules such as CD31 on resting T cells in circulation and their ligand on endothelium, resulting in preferential activation of VLA-4 mediated adhesion; and 3) VLA-4 binding to VCAM-1 on endothelium, resulting in strong adhesion. Subsequent events would then mediate migration into tissue.

For T cell binding to APC, a prototypic sequence of events for a successful antigen-specific interaction would be: 1) weak contact between migrating T cell in tissue and the APC (whose molecular basis is not yet well defined); 2) CD3/TCR engagement of MHC/peptide on the opposing cell, inducing preferentially LFA-1 integrin function; 3) LFA-1-mediated adhesion to ICAM-1, ICAM-2 or ICAM-3 on the APC. Trigger molecules such as CD7 and CD28 likely contribute to adhesion regulation during T cell interaction with certain cells,[28] but where they fit temporally in the process remains to be defined. In addition, the subsequent activation of the T cell is made possible by signaling properties of the adhesion molecules.

REFERENCES

1. M.E. Hemler, Structures and Functions of VLA proteins and related integrins, In: *Receptors for Extracellular Matrix*, edited by Mecham, R.P. and McDonald, J.A. San Diego: Academic Press, p. 255 (1991)
2. R. Rothlein, M. Czajkowski and T.K. Kishimoto, Intercellular adhesion molecule-1 (ICAM-1) in the inflammotory response, *in preparation* (1991).
3. T.A. Springer, Adhesion receptors of the immune system, *Nature* 346:425 (1990).
4. T.D. Geppert, L.S. Davis, H. Gur, M.C. Wacholtz and P.E. Lipsky, Accessory cell signals involved in T-cell activation, *Immunol.Rev.* 117:5 (1990).
5. E.L. Berg, L.J. Picker, M.K. Robinson, P.R. Streeter and E.C. Butcher, Vascular Addressins: Tissue Selective Endothelial Cell Adhesion Molecules for Lymphocyte Homing, In: *Cellular and Molecular Mechanisms of Inflammation, Volume 2*, edited by : Academic Press, p. 111 (1991)
6. Y. Shimizu and S. Shaw, Lymphocyte interactions with extracellular matrix, *FASEB J.* 5:2292 (1991).
7. Y. Tanaka, S. Shaw and K.J. Horgan, Human naive and memory T cells, *Med. Immunol.* 21:527 (1991).

8. T. Schweighoffer and S. Shaw, Regulation of adhesiveness: mechanisms of expression, activation and shedding, In: *Handbook of Immunopharmacology: Adhesion molecules*, edited by Wegner, C.D. London: Academic Press, (1992)

9. Y. Shimizu, W. Newman, Y. Tanaka and S. Shaw, Lymphocyte/endothelial interactions, *Immunol. Today* in press:(1992).

10. E.C. Butcher, Leukocyte-endothelial cell recognition: three (or more) steps to specificity and diversity, *Cell* 67:1033 (1991).

11. D.R. Phillips, I.F. Charo and R.M. Scarborough, GPIIb-IIIa: The responsive integrin, *Cell* 65(3):359 (1991).

12. N. Kieffer and D.R. Phillips, Platelet membrane glycoproteins: functions in cellular interactions, *Annu. Rev. Cell Biol.* 6:329 (1990).

13. M.L. Dustin, J. Garcia-Aguilar, M.L. Hibbs, R.S. Larson, S.A. Stacker, D.E. Staunton, A.J. Wardlaw and T.A. Springer, Structure and regulation of the leukocyte adhesion receptor LFA-1 and its counterreceptors, ICAM-1 and ICAM-2, *Cold Spring Harbor Symp. Quant. Biol.* 54:753 (1989).

14. A.R. De Fougerolles and T.A. Springer, Intercellular adhesion molecule 3, a third adhesion counter-receptor for lymphocyte function-associated molecule 1 on resting lymphocytes, *J. Exp. Med.* 175:185 (1992).

15. A.R. De Fougerolles, S.A. Stacker, R. Schwarting and T.A. Springer, Characterization of ICAM-2 and evidence for a third counter-receptor for LFA-1, *J. Exp. Med.* 174:253 (1991).

16. Y. Shimizu, W. Newman, N. Graber, K.J. Horgan, L.D. Beall, T.V. Gopal, G.A. van Seventer and S. Shaw, Four molecular pathways of T cell adhesion to endothelial cells: roles of LFA-1, VCAM-1 and ELAM-1 and changes in pathway hierarchy under different activation conditions, *J. Cell Biol.* 113:1203 (1991).

17. M.A. Arnaout, Leukocyte adhesion molecules deficiency: its structural basis, pathophysiology and implications for modulating the inflammatory response, *Immunol. Rev.* 114:145 (1990).

18. J.L. Bednarczyk and B.W. McIntyre, A monoclonal antibody to VLA-4 α-chain (CDw49d) induces homotypic lymphocyte aggregation, *J. Immunol.* 144(3):777 (1990).

19. M.R. Campanero, R. Pulido, M.A. Ursa, M. Rodriguez-Moya, M.O. De Landazuri and F. Sanchez-Madrid, An alternative leukocyte homotypic adhesion mechanism, LFA-1/ICAM-1 independent, triggered through the human VLA-4 integrin, *J. Cell Biol.* 110:2157 (1990).

20. L. Osborn, C. Hession, R. Tizard, C. Vassallo, S. Luhowskyj, G. Chi-Rosso and R. Lobb, Direct expression cloning of vascular cell adhesion molecule 1, a cytokine-induced endothelial protein that binds to lymphocytes, *Cell* 59:1203 (1989).

21. T.B. Issekutz and A. Wykretowicz, Effect of a new monoclonal antibody, TA-2, that inhibits lymphocyte adherence to cytokine stimulated endothelium in the rat, *J. Immunol.* 147:109 (1991).

22. T.B. Issekutz, Inhibition of in vivo lymphocyte migration to inflammation and homing to lymphoid tissues by the TA-2 monoclonal antibody: a likely role for VLA-4 in vivo, *J. Immunol.* 147:4178 (1991).

23. G.E. Rice, J.M. Munro, C. Corless and M.P. Bevilacqua, Vascular and nonvascular expression of INCAM-110. A target for mononuclear leukocyte adhesion in normal and inflamed human tissues, *Am. J. Pathol.* 138:385 (1991).

24. G.A. van Seventer, W. Newman, Y. Shimizu, T.B. Nutman, Y. Tanaka, K.J. Horgan, T.V. Gopal, E. Ennis, D. O'Sullivan, H. Grey and S. Shaw, Analysis of T-cell stimulation by superantigen plus MHC class II molecules or by CD3 mAb: costimulation by purified adhesion ligands VCAM-1, ICAM-1 but not ELAM-1, *J. Exp. Med.* 174:901 (1991).

25. M.L. Dustin and T.A. Springer, T-cell receptor cross-linking transiently stimulates adhesiveness through LFA-1, *Nature* 341:619 (1989).

26. Y. Shimizu, G.A. van Seventer, K.J. Horgan and S. Shaw, Roles of adhesion molecules in T cell recognition: Fundamental similarities between four integrins on resting human T cells (LFA-1, VLA-4, VLA-5, VLA-6) in expression, binding, and costimulation, *Immunol. Rev.* 114:109 (1990).

27. B.M.C. Chan, J.G.P. Wong, A. Rao and M.E. Hemler, T cell receptor-dependent, antigen-specific stimulation of a murine T cell clone induces a transient, VLA protein-mediated binding to extracellular matrix, *J. Immunol.* 147:398 (1991).

28. Y. Shimizu, G.A. van Seventer, E. Ennis, W. Newman, K.J. Horgan and S. Shaw, Crosslinking of the T cell-specific accessory molecules CD7 and CD28 modulates T cell adhesion, *J. Exp. Med.* 175:577 (1992).

29. G. Koopman, Y. van Kooyk, M. De Graaff, C.J.L.M. Meyer, C.G. Figdor and S.T. Pals, Triggering of the CD44 antigen on T lymphocytes promotes T cell adhesion through the LFA-1 pathway, *J. Immunol.* 145:3589 (1990).

30. Y. Tanaka, S.M. Albelda, K.J. Horgan, G.A. van Seventer, Y. Shimizu, W. Newman, J. Hallam, P.J. Newman, C.A. Buck and S. Shaw, CD31/PECAM-1 is a preferential amplifier of VLA-integrin-mediated adhesion for distinctive T cell subsets, *submitted* (1992).

31. W.A. Muller, C.M. Ratti, S.L. McDonnel and Z.A. Cohn, A human endothelial cell-restricted externally disposed plasmalemmal protein enriched in intracellular junctions, *J. Exp. Med.* 170:399 (1989).

32. D. Guy-Grand, N. Cerf-Bensussan, B. Malissen, M. Malassis-Seris, C. Briottet and P. Vassalli, Two gut intraepithelial CD8$^+$ lymphocyte populations with different T cell receptors: A role for the gut epithelium in T cell differentiation, *J. Exp. Med.* 173:471 (1991).

33. B. Holzmann, B.W. McIntyre and I.L. Weissman, Identification of a murine Peyer's patch-specific lymphocyte homing receptor as an integrin molecule with an alpha chain homologous to human VLA-4, *Cell* 56:37 (1989).

34. B. Holzmann and I.L. Weissman, Peyer's patch-specific lymphocyte homing receptors consist of a VLA-4-like alpha chain associated with either of two integrin beta chains, one of which is novel, *EMBO J.* 8:1735 (1989).

35. C.R. Mackay, W.L. Marston and L. Dudler, Naive and memory T cells show distinct pathways of lymphocyte recirculation, *J. Exp. Med.* 171:801 (1990).

36. S.M. Albelda, W.A. Muller, C.A. Buck and P.J. Newman, Molecular and cellular properties of PECAM-1 (endoCAM/CD31): a novel vascular cell-cell adhesion molecule, *J. Cell Biol.* 114:1059 (1991).

37. A.F. Williams, Out of equilibrium, *Nature* 352:473 (1991).

38. O. Spertini, F.W. Luscinskas, G.S. Kansas, J.M. Munro, J.D. Griffin, M.A. Gimbrone, Jr. and T.F. Tedder, Leukocyte adhesion molecule-1 (LAM-1, L-selectin) interacts with an inducible endothelial cell ligand to support leukocyte adhesion, *J. Immunol.* 147:2565 (1991).

39. W.M. Gallatin, I.L. Weissman and E.C. Butcher, A cell surface molecule involved in organ-specific homing of lymphocytes, *Nature* 304:30 (1983).

40. A. Hamann, D. Jablonski-Westrich, P. Jonas and H.G. Thiele, Homing receptors reexamined: mouse LECAM-1 (MEL-14 antigen) is involved in lymphocyte migration into gut-associated lymphoid tissue, *Eur. J. Immunol.* 21:2925 (1991).

41. M.W. Makgoba, M.E. Sanders and S. Shaw, The CD2-LFA-3 and LFA-1-ICAM-1 pathways: relevance to T-cell recognition, *Immunol. Today* 10:417 (1989).

ANALYSES OF VLA-4 STRUCTURE AND FUNCTION

Paul D. Kassner[1], Joaquin Teixidó[1], Bosco M.C. Chan[1], Christina M. Parker[2] and Martin E. Hemler[1]

[1]Division of Tumor Virology
[2]Division of Immunochemistry, Department of Rheumatology and Immunology
Dana-Farber Cancer Institute
Harvard Medical School
Boston, MA 02115

INTRODUCTION

The integrin VLA-4 mediates the adhesion of lymphocytes, monocytes, eosinophils and some melanoma cells to VCAM-1 on the surface of activated endothelial cells.[1-7] Because the VLA-4/VCAM-1 pathway plays a prominent role in the migration and localization of these various cell types, it is a potential target of therapeutic intervention that has relevance to inflammation, atherosclerosis,[8] allergy and asthma,[9,10] and possibly also tumor cell metastasis[11] and arthritis.[12]

The VLA-4 heterodimer also mediates cell attachment to the alternatively spliced "CS1" region in fibronectin.[13-15] The VLA-4/fibronectin interaction i) may play a key role in the localization of antigen-specific lymphocytes to sites of antigen challenge,[16] ii) can facilitate the costimulation of T lymphocytes,[17-19] iii) may have important functional relevance during maturation of bone marrow progenitor cells,[20] and iv) may be important for neural crest cell migration during embryogenesis.[21]

In addition, VLA-4 plays a role in the triggering of homotypic aggregation among various lymphoid and myeloid cell lines.[22,23] This latter activity does not appear to involve either fibronectin or VCAM-1.[24] Interestingly, the diverse functions of VLA-4 may involve distinct but overlapping epitopes within the VLA-4 molecule.[3,24]

Compared to other integrins, the VLA-4 molecule has some unique structural features. Most notably, the α^4 subunit of VLA-4 is the only integrin α subunit known to undergo variable proteolytic cleavage near the middle of the molecule, yielding fragments of 70 and 80 kD.[5] This cleavage event is increased on T-lymphocytes following activation,[13,25,26] and it has been hypothesized that different forms of the molecule may have different functional capabilities.[13,27]

The primary amino acid sequence of α^4 subunit fails to align closely with any of the other known integrin α subunit sequences.[5,28,29] As shown in Figure 1, the cytoplasmic tail sequence of human α^4 is nearly identical to that of mouse α^4 (in 32/33 residues), suggesting that this sequence may have an important function which is conserved across species. In contrast, there are pronounced differences between α^4 and the various other integrin α chain cytoplasmic tails, suggesting that the function of the α^4 cytoplasmic tail could be unique.

In this paper we discuss experiments in which we have identified the proteolytic cleavage

Mechanisms of Lymphocyte Activation and Immune Regulation IV: Cellular Communications
Edited by S. Gupta and T.A. Waldmann, Plenum Press, New York, 1992

163

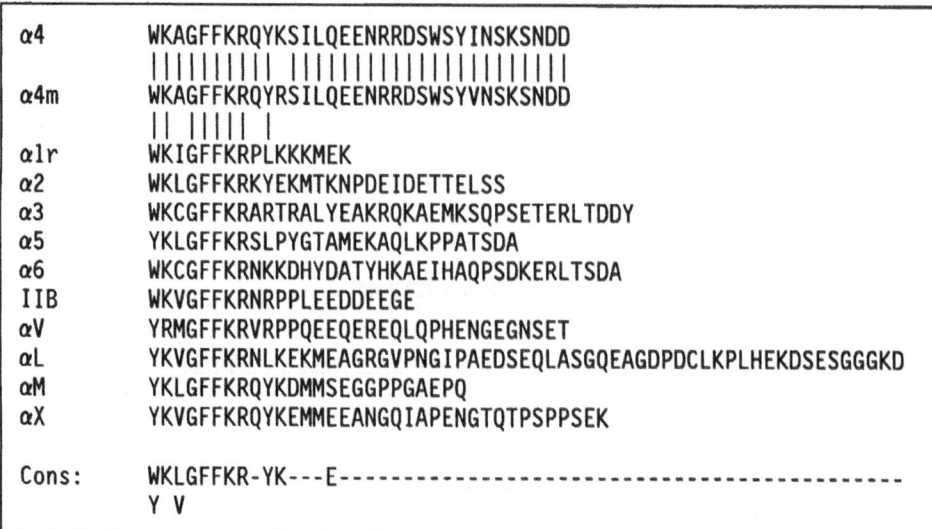

```
α4      WKAGFFKRQYKSILQEENRRDSWSYINSKSNDD
        |||||||||| |||||||||||||||||||||||
α4m     WKAGFFKRQYRSILQEENRRDSWSYVNSKSNDD
        || |||||| |
αlr     WKIGFFKRPLKKKMEK
α2      WKLGFFKRKYEKMTKNPDEIDETTELSS
α3      WKCGFFKRARTRALYEAKRQKAEMKSQPSETERLTDDY
α5      YKLGFFKRSLPYGTAMEKAQLKPPATSDA
α6      WKCGFFKRNKKDHYDATYHKAEIHAQPSDKERLTSDA
IIB     WKVGFFKRNRPPLEEDDEEGE
αV      YRMGFFKRVRPPQEEQEREQLQPHENGEGNSET
αL      YKVGFFKRNLKEKMEAGRGVPNGIPAEDSEQLASGQEAGDPDCLKPLHEKDSESGGGKD
αM      YKLGFFKRQYKDMMSEGGPPGAEPQ
αX      YKVGFFKRQYKEMMEEANGQIAPENGTQTPSPPSEK

Cons:   WKLGFFKR-YK---E-----------------------------------------
        Y V
```

Figure 1. Comparison of integrin α chain cytoplasmic domain sequences. All sequences are for human α chains except for the α[4] sequence from mouse[30] and the α[1] sequence obtained from rat. The α[6] sequence is from Tamura et al.,[31] and references for other sequences are listed in a recent review article.[29] The indicated consensus residues are those appearing in 4 or more of the 11 listed sequences.

site within the α[4] subunit, used site-directed mutagenesis to abolish cleavage, and made functional comparisons between cells expressing cleaved and uncleaved α[4] subunits. Also, we discuss other experiments in which we utilized chimeric α subunit constructs to analyze the functional role of the integrin α[4] cytoplasmic tail.

FUNCTIONAL CONSEQUENCES OF α[4] CLEAVAGE

Our recent investigations of the cleavage of the 150 kD α[4] subunit into 70 and 80 kD fragments have suggested that this is a regulated, compartmentalized event, rather than an artifact of cell lysis.[32] To identify the exact location of the α[4] cleavage site, we purified a substantial quantity of the 70 kD fragment (which is C-terminal to the 80 kD fragment) and determined its amino–terminal sequence. As shown in Figure 2, the amino acid sequence obtained from the 70 kD fragment (STEEFPPLQP) exactly matched the primary amino acid sequence determined from α[4] cDNA,[28] from Ser (559) through Pro (567). This result is consistent with cleavage of α[4] occurring just after the dibasic residues Lys (557) and Arg (558). Site directed mutagenesis experiments were then carried out to replace either Lys (557) or Arg (558) with either Leu or Gln as indicated in Figure 2. Also, in a control experiment, another mutation was made, replacing both Lys (574) and Glu (575) with an Asn-Ala sequence. After transfection of these various constructs into the cell line K-562, cells were selected to obtain comparable cell surface expression. Then, each transfected cell line was analyzed for α[4] subunit cleavage, adhesion to sVCAM, and adhesion to FN-40. As shown in Table 1, a high degree of cleavage was observed for unmutated α[4] cells, and in 4KE/NA-transfected cells, but cleavage was completely abolished in cells transfected with the 4R/L construct, and almost completely abolished in 4K/Q cells. These results confirm that the dibasic residues Lys (557) and Arg (558) immediately preceding the cleavage point are critical for recognition by the relevant protease. Although the protease has not been identified, we surmise that it could belong to the rapidly growing subtilisin–like family of proteases that cleave following dibasic residues[34].

Although cleavage was abolished or nearly abolished in two of four transfectants, there was no correlation between α[4] cleavage and the ability of cells to adhere to either fibronectin or

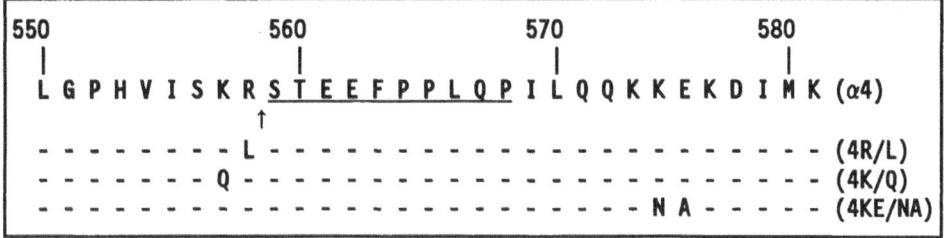

Figure 2. Site–directed mutations made in the vicinity of the α^4 cleavage site (indicated by ↑). The residues identified from N–terminal amino acid sequencing of the α^4 70 kD cleavage fragment are underlined. Also, the locations of three separate site–directed mutations are indicated.

Table 1. Structural and functional consequences of mutations in the vicinity of the α^4 cleavage site.

cDNA Transfected into K562 cells	80/70 Cleavage[a]	Adhesion[b] to FN-40	Adhesion[b] to sVCAM-1	α^4-triggered[c] aggregation
--	N/A	12 ± 5	4 ± 0.2	No
α^4	90-99%	96 ± 11	585 ± 140	Yes
4KE/NA	90-99%	172 ± 65	609 ± 6	Yes
4R/L	0%	103 ± 26	478 ± 62	Yes
4K/Q	< 2%	124 ± 35	515 ± 7	Yes

[a]Cleavage was estimated visually following immunoprecipitation of ^{125}I-labeled α^4.
[b]Adhesion assays utilized 96-well plastic plates coated with the 40 kD chymotryptic fragment of fibronectin (at 0.4 μg/ml) or with a soluble form of VCAM-1 (at 0.3 μg/ml). Values for cells bound/mm^2 were obtained as described elsewhere[32,33].
[c]The MAb HP1/7 was used to initiate α^4-triggered aggregation as previously described[23,24].

sVCAM. Regardless of the type of α^4 mutation, each transfected cell adhered similarly to the 40 kD fragment of fibronectin (which contains the CS1 region) and to the VLA-4 ligand VCAM-1. Also, there was no detectable difference in the ability of the transfected cells to aggregate in reponse to triggering by anti–α^4 antibodies (Table 1).

In conclusion, despite previous speculation to the contrary, we have found that α^4 subunit cleavage has no apparent effects on known VLA-4 adhesion and aggregation functions.

FUNCTIONAL ROLE OF THE α^4 CYTOPLASMIC TAIL

The prediction from the data in Figure 1 is that the cytoplasmic tail of α^4 may have a unique function. Thus, to directly analyze the functional role of the α^4 cytoplasmic tail, two sets of chimeric molecules were prepared. As shown schematically in Figure 3, we prepared molecules with extracellular and transmembrane domains from α^4 (the X4 series) and cytoplasmic domains from either α^4, α^2, or α^5, yielding constructs named X4C4, X4C2, and X4C5 respectively. A similar set of chimeric molecules containing the extracellular and transmembrane portions of α^2 was also prepared (called X2C2, X2C4, X2C5).

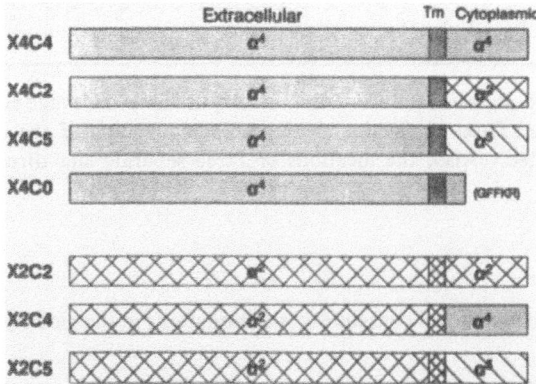

Figure 3. Schematic diagram of integrin α chain chimeric constructs. To prepare the chimeric cDNA constructs, a common restriction site was created by site–directed mutagenesis just after the last amino acid of the transmembrane region as described elsewhere[33].

Analysis of the X4 Series of Chimeras

First, the three constructs in the X4 series were transfected into a VLA–4–negative cell line K562 (in the erythroid lineage). Then, after selection to obtain near equivalent expression, transfected K562 cells were analyzed for their contributions to VLA–4 adhesion functions. As shown in Table 2, all three constructs gave similar levels of adhesion to plastic surfaces coated with the VLA–4 ligands FN–40 and VCAM–1. In contrast, mock–transfected K562 cells adhered relatively poorly to the same ligands. In conclusion, it appears that the cytoplasmic tail of α^4 does not specifically contribute to VLA–4 adhesion or aggregation functions, since it can be replaced by other cytoplasmic tails. Also, K562 cells transfected with any of the three constructs were readily triggered to aggregate in response to either anti–α^4 or anti–β_1 antibodies. Thus, again the cytoplasmic tail of α^4 did not appear to play a specific role.

Table 2. Effect of replacement of the α^4 cytoplasmic tail on VLA–4 adhesion functions

Transfected K562 Cells	Adhesion to FN-40[a]	Adhesion to VCAM-1[a]	α^4-triggered aggregation[b]	β_1-triggered aggregation[b]
KpFNEO	159 ± 14	53 ± 23	-	-
KX4C2	689 ± 29	822 ± 48	+++	+++
KX4C4	720 ± 57	890 ± 55	+++	+++
KX4C5	741 ± 43	834 ± 49	+++	+++

[a]Adhesion (reported as cells bound/mm^2) was determined using either FN–40 or VCAM–1 coated onto plastic at 40 µg/ml or 5 µg/ml as described elsewhere.[32,35]
[b]Aggregation was triggered by MAb to either α^4 or β_1 as previously described[24,36] and was assessed visually with the aid of a microscope.

To analyze the functional contributions of the cytoplasmic tail of α^4 in another context, constructs in the X2 series (Figure 3) were utilized. Because it lacked endogenous VLA-2 expression, the rhabdomyosarcoma cell line RD was chosen as recipient for the X2C2, X2C4, and X2C5 constructs. After transfection and selection for near equivalent expression, adhesion

Table 3. Effect of replacement of α^2 cytoplasmic tail on VLA-2 adhesion functions.

Transfected RD cells	Adhesion to Collagen[a]	Adhesion to to Laminin	Adhesion to Fibronectin
RDpF	126 ± 26	94 ± 25	990 ± 78
RDX2C2	523 ± 26	448 ± 46	1043 ± 34
RDX2C4	587 ± 49	516 ± 50	1040 ± 64
RDX2C5	614 ± 28	576 ± 31	949 ± 12

[a]Adhesion (reported as cells bound/mm^2) was determined using 96-well plastic plates coated with either collagen (1.3 μg/ml), mouse laminin (2.5 μg/ml) or fibronectin (5.0 μg/ml) as described elsewhere.[32,33]

to established VLA-2 ligands was analyzed (Table 3). As shown, RD cells transfected with each construct yielded similar levels of adhesion to either collagen or laminin. In contrast, mock-transfected RD cells yielded only a low background level of adhesion, most likely mediated by other integrins (VLA-1, VLA-6) endogenously expressed on RD cells. In a control experiment, both mock-transfected and transfected RD cells adhered similarly to fibronectin, consistent with previous findings that VLA-2 is not a receptor for fibronectin.[37,38]

Although cell adhesion was apparently unaffected by swapping of cytoplasmic tails, we hypothesized that cell migration might be more likely to require a specific interaction between integrin α chain cytoplasmic tails and the underlying cytoskeletal machinery. In a random cell migration assay, the various RD cell transfectants were allowed to adhere to collagen, laminin,

Table 4. Effect of replacement of α^2 cytoplasmic tail on VLA-2-dependent random cell migration.

Transfected RD cells	Migration on Collagen[a]	Migration on Laminin	Migration on Fibronectin
RDpF	10.7 ± 6.7	10.8 ± 6.3	27.0 ± 13.0
RDX2C2	11 ± 8.5	11.5 ± 6.5	28.0 ± 11.9
RDX2C4	36.8 ± 11.8	27.8 ± 12.3	27.6 ± 11.9
RDX2C5	7.9 ± 6.7	12.2 ± 7.8	25.5 ± 11.0

[a]Random cell migration was carried out using the indicated transfected RD cells on collagen, laminin, or fibronectin each coated onto plastic at 1 μg/ml as previously described. For each experiment, results are reported ± 1 S.D. and N = 50–60. The migration rates indicated by boxes are significantly greater than the migration rates of the other cells on the same ligand (p < 0.001).

or fibronectin, and then the position of individual cells was recorded every hour for four hours. As shown in Table 4, RD cells transfected with the X2C4 construct showed significantly greater migration rates than the other cells on both collagen and laminin. Importantly, the X2C2 and X2C5 transfectants showed migration rates very similar to those of mock transfected RD cells (RDpF), suggesting that the C2 and C5 sequences were not contributing a negative migration signal. In a control experiment, there was no appreciable difference in the migration rates on fibronectin, as expected since VLA-2 does not mediate adhesion to fibronectin.

Together these results indicate that the cytoplasmic domain of α^4 confers on VLA-2 the unique ability to mediate increased cell migration on collagen and laminin. Notably, this occurred even though there were no detectable differences in cell adhesion to collagen and laminin. Thus, the cytoplasmic domain of α^4 appears to be critically involved in a post-ligand binding event, whereby the information derived from cell adhesion is specifically translated into cell migration. In future studies it will be important to further examine the role of the α^4 cytoplasmic domain, to determine whether it supports migration in the context of the intact VLA-4 molecule.

Because all of the known integrin α subunits have cytoplasmic tails that are quite diverse (See Figure 1), we hypothesize that each integrin may be specifically suited to carry out diverse post-ligand binding events. Thus, even though there are several integrins that act as apparently redundant receptors (eg. for fibronectin, laminin, and collagen), ligand binding to each of these different receptors could conceivably result in a diversity of subsequent events.

ACKNOWLEDGEMENTS

This work was supported by National Institutes of Health grants GM46526, GM38903 and CA42368 (to M.E.H.), a Centennial Fellowship from the Medical Research Council of Canada (to B.M.C.C.) and an NIH Physician Scientist Training Award (K11 AI00903, to C.M.P.). We acknowledge H. Randolph Byers (Mass. General Hospital, Boston, MA) for assistance with cell migration assays, and The American Red Cross Blood Services–Northeast Region, for providing human plasma for use in preparation of fibronectin fragments.

REFERENCES

1. L. Osborn, C. Hession, R. Tizard, C. Vasallo, S. Luhowskyj, G. Chi-Rosso,and R. Lobb, Direct cloning of vascular cell adhesion molecule 1 (VCAM1), a cytokine-induced endothelial protein that binds to lymphocytes, *Cell* 59:1203 (1989).
2. G.E. Rice, J.M. Munro, and M.P. Bevilacqua, Inducible cell adhesion molecule 110 (INCAM-110) is an endothelial receptor for lymphocytes: A CD11/CD18-independent adhesion mechanism, *J. Exp. Med.* 171:1369 (1990).
3. M.J. Elices, L. Osborn, Y. Takada, C. Crouse, S. Luhowskyj, M.E. Hemler, and R.R. Lobb, VCAM-1 on activated endothelium interacts with the leukocyte integrin VLA-4 at a site distinct from the VLA-4/fibronectin binding site, *Cell* 60:577 (1990).
4. B.R. Schwartz, E.A. Wayner, T.M. Carlos, H.D. Ochs, and J.M. Harlan, Identification of surface proteins mediating adherence of CD11/CD18-deficient lymphoblastoid cells to cultured human endothelium, *J. Clin. Invest.* 85:2019 (1990).
5. M.E. Hemler, M.J. Elices, C. Parker, and Y. Takada, Structure of the integrin VLA-4 and its cell-cell and cell-matrix adhesion functions, *Immunol. Rev.* 114:45 (1990).
6. P. Allavena, C. Paganin, I. Martin-Padura, G. Peri, M. Gaboli, E. Dejana, P.C. Marchisio, and A. Mantovani, Molecules and structures involved in the adhesion of natural killer cells to vascular endothelium, *J. Exp. Med.* 173:439 (1991).
7. T.B. Issekutz and A. Wyrkretowicz, Effect of a new monoclonal antibody, TA-2, that inhibits lymphocyte adherence to cytokine stimulated endothelium in the rat, *J. Immunol.* 147:109 (1991).

8. A. Laffón, R. García-Vincuña, A. Humbría, A.A. Postigo, A.L. Corbí, M.O. De Landázuri, and F. Sánchez-Madrid, Upregulated expression and function of VLA-4 fibronectin receptors on human activated T cells in rheumatoid arthritis, *J. Clin. Invest.* 88:546 (1991).

9. G.M. Walsh, J-J. Mermod, A. Hartnell, A.B. Kay, and A.J. Wardlaw, Human eosinophil, but not neutrophil, adherence to IL-1-stimulated human umbilical vascular endothelial cells is $\alpha_4\beta_1$ (very late antigen-4) dependent, *J. Immunol.* 146:3419 (1991).

10. B.S. Bochner, F.W. Luscinskas, M.A. Gimbrone, W. Newman, S.A. Sterbinsky, C.P. Derse-Anthony, D. Klunk, and R.P. Schleimer, Adhesion of human basophils, eosinophils, and neutrophils to interleukin 1-activated human vascular endothelial cells: contributions of endothelial cell adhesion molecules, *J. Exp. Med.* 173:1553 (1991).

11. G.E. Rice and M.P. Bevilacqua, An inducible endothelial cell surface glycoprotein mediates melanoma adhesion, *Science* 246:1303 (1989).

12. M.I. Cybulsky and M.A. Gimbrone, Endothelial expression of a mononuclear leukocyte adhesion molecule during atherogenesis, *Science* 251:788 (1991).

13. E.A. Wayner, A. García-Pardo, M.J. Humphries, J.A. McDonald, and W.G. Carter, Identification and characterization of the lymphocyte adhesion receptor for an alternative cell attachment domain in plasma fibronectin, *J. Cell Biol.* 109:1321 (1989).

14. J.-L. Guan and R.O. Hynes, Lymphoid cells recognize an alternatively spliced segment of fibronectin via the integrin receptor $\alpha4\beta1$, *Cell* 60:53 (1990).

15. A. García-Pardo, E.A. Wayner, W.G. Carter, and O.C. Ferreira, Human B lymphocytes define an alternative mechanism of adhesion to fibronectin: The interaction of the $\alpha4\beta1$ integrin with the LHGPEILDVPST sequence of the type III connecting segment is sufficient to promote cell attachment, *J. Immunol.* 144:3361 (1990).

16. T.A. Ferguson, H. Mizutani, and T.S. Kupper, The integrin-binding peptides GRGDSP and GPEILDVPST abrogate T cell mediated immune responses in vivo, *Proc. Natl. Acad. Sci. USA* (1991).

17. L.S. Davis, N. Oppenheimer-Marks, J.L. Bednarczyk, B.W. McIntyre, and P.E. Lipsky, Fibronectin promotes proliferation of naive and memory T cells by signaling through both the VLA-4 and VLA-5 integrin molecules, *J. Immunol.* 145:785 (1990).

18. Y. Nojima, M.J. Humphries, A.P. Mould, A. Komoriya, K.M. Yamada, S.F. Schlossman, and C. Morimoto, VLA-4 mediates CD3-dependent CD4+ T cell activation via the CS1 alternatively spliced domain of fibronectin, *J. Exp. Med.* 172:1185 (1990).

19. Y. Shimizu, G.A. Van Seventer, K.J. Horgan, and S. Shaw, Costimulation of proliferative responses of resting CD4+ T cells by the interaction of VLA-4 and VLA-5 with fibronectin or VLA-6 with laminin, *J. Immunol.* 145:59 (1990).

20. D.A. Williams, M. Rios, C. Stephens, and V.P. Patel, Fibronectin and VLA-4 in haematopoietic stem cell-microenvironment interactions, *Nature* 352:438 (1991).

21. S. Dufour, J-L. Duband, M.J. Humphries, M. Obara, K.M. Yamada, and J.P. Thiery, Attachment, spreading and locomotion of avian neural crest cells are mediated by multiple adhesion sites on fibronectin molecules, *EMBO J.* 7:2661 (1988).

22. J.L. Bednarczyk and B.W. McIntyre, A monoclonal antibody to VLA-4 α chain (CDw49) induces homotypic lymphocyte aggregation, *J. Immunol.* 144:777 (1990).

23. M.R. Campanero, R. Pulido, M.A. Ursa, M. Rodriquez-Moya, M.O. De Landazuri, and F. Sánchez-Madrid, An alternative leukocyte homotypic adhesion mechanism, LFA-1/ICAM-1 independent, triggered through the human VLA-4 integrin, *J. Cell. Biol.* 110:2157 (1990).

24. R. Pulido, M.J. Elices, M.R. Campanero, L. Osborn, S. Schiffer, A. García-Pardo, R. Lobb, M.E. Hemler, and F. Sánchez-Madrid, Functional evidence for three distinct and independently inhibitable adhesion activities mediated by the human integrin VLA-4: Correlation with distinct α4 epitopes, *J. Biol. Chem.* 266:10241 (1991).

25. F. Sánchez-Madrid, M.O. De Landazuri, G. Morago, M. Cebrian, A. Acevedo, and C. Bernabeu, VLA-3: a novel polypeptide association within the VLA molecular complex: cell distribution and biochemical characterization, *Eur. J. Immunol.* 16:1343 (1986).

26. B.W. McIntyre, E.L. Evans, and J.L. Bednarczyk, Lymphocyte surface antigen L25 is a member of the integrin receptor superfamily, *J. Biol. Chem.* 264:13745 (1989).

27. J.L. Bednarczyk, J.N. Wygant, and B.W. McIntyre, Structural and functional analysis of the integrin VLA-4, *J. Cell. Biochem.* 14A:184 (1990).

28. Y. Takada, M.J. Elices, C. Crouse, and M.E. Hemler, The primary structure of the $\alpha4$ subunit of VLA-4: homology to other integrins and a possible cell-cell adhesion function, *EMBO J.* 8:1361 (1989).

29. M.E. Hemler, Structures and functions of VLA proteins and related integrins, in: "Receptors for extracellular matrix proteins (A volume of Biology of Extracellular Matrix)", R.P. Mecham et al., eds., Academic Press, Inc., San Diego, CA, pp. 255-299 (1991).

30. H. Neuhaus, M.C-T. Hu, M.E. Hemler, Y. Takada, B. Holzmann, and I.L. Weissman, Cloning and expression of cDNAs for the α subunit of the murine lymphocyte-peyer's patch homing receptor, *J. Cell Biol.* 115:1149 (1991).

31. R.N. Tamura, C. Rozzo, L. Starr, J. Chambers, L.F. Reichardt, H.M. Cooper, and V. Quaranta, Epiththelial integrin $\alpha6\beta4$: complete primary structure of $\alpha6$ and variant forms of $\beta4$, *J. Cell Biol.* 111:1593 (1990).

32. J. Teixidó, C.M. Parker, P.D. Kassner, and M.E. Hemler, Functional and structural analysis of VLA-4 integrin α^4 subunit cleavage, *J. Biol. Chem.* 267:1786 (1992).

33. B.M.C. Chan, P.D. Kassner, J.A. Schiro, H.R. Byers, T.S. Kupper, and M.E. Hemler, Distinct cellular functions mediated by different VLA integrin α subunit cytoplasmic domains, *Cell* 68:1051 (1992).

34. P.J. Barr, Mammalian subtilisins: the long-sought dibasic processing endoproteases, *Cell* 66:1 (1991).

35. B.M.C. Chan, M.J. Elices, E. Murphy, and M.E. Hemler, Adhesion to VCAM-1 and fibronectin: comparison of $\alpha4\beta1$ (VLA-4) and $\alpha4\beta7$ on the human cell line JY, *J. Biol. Chem.* (in press, 1992).

36. M.R. Campanero, A.G. Arroyo, R. Pulido, A. Ursa, M.S. de Matías, P. Sánchez-Mateos, P.D. Kassner, B.M.C. Chan, M.E. Hemler, A. Corbí, M.O. De Landázuri, and F. Sánchez-Madrid, Triggering of integrin-mediated intercellular adhesion through the common $\beta1$ subunit of VLA heterodimers: Evidence for a role of $\alpha2/\beta1$ and $\alpha4/\beta1$ in cell-cell interactions, (submitted, 1992).

37. M.J. Elices and M.E. Hemler, The human integrin VLA-2 is a collagen receptor on some cells and a collagen/laminin receptor on others, *Proc. Natl. Acad. Sci. USA* 86:9906 (1989).

38. B.M.C. Chan, N. Matsuura, Y. Takada, B.R. Zetter, and M.E. Hemler, In vitro and in vivo consequences of VLA-2 expression on rhabdomyosarcoma cells, *Science* 251:1600 (1991).

ON THE REGULATION OF β2 INTEGRINS

M. Amin Arnaout, Masahiro Michishita, and Chander P. Sharma

Leukocyte Biology and Inflammation Program
Renal Unit and Department of Medicine
Massachusetts General Hospital and
Harvard Medical School, Boston, MA

INTRODUCTION

β2 integrins (Leu-CAMs, CD11/CD18) are three heterodimeric surface membrane glycoprotein receptors which mediate a large number of divalent-cation-dependent cell-cell and cell-matrix adhesion functions in leukocytes.[1] Each heterodimer (Figure 1) consists of a distinct α subunit (CD11a, CD11b or CD11c) noncovalently associated with a single β subunit (CD18). CD11a/CD18 is expressed on all leukocytes. CD11b and CD11c expression is restricted to myelomonocytic cells and NK cells.

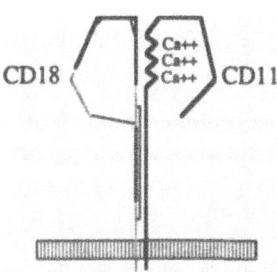

Figure 1. Schematic of the jelly-fish like structure of β2 integrins drawn to scale. The schematic is modified from,[2] based on the EM structure of other integrins[3] and knowledge of the cysteine-bridges in the homologous β3 subunit[4] The dimensions of integrins as described for the β1 integrin α5β1[3] are: 12-15 x 8 nm for the ovoid head, 12-15 nm for the remainder of the C-terminal portion of each subunit, and 2 nm distance between the tails, each of which spans the plasma membrane once. Each of the CD11 subunits contains three divalent-cation binding sites. The portion of CD18 outlined in black corresponds to the highly conserved 250 amino acid segment involved in formation of the heterodimer[5] and perhaps in ligand binding . The relative position of the four cysteine-rich repeats

Mechanisms of Lymphocyte Activation and Immune Regulation IV: Cellular Communications
Edited by S. Gupta and T.A. Waldmann, Plenum Press, New York, 1992

171

in CD18 is outlined by four contiguous open rectangles. The four major cysteine-bridges in CD18 are also shown (thin stippled lines).

Molecular cloning revealed similarities in the overall structure of the CD11 subunits with nonconserved short cytoplasmic tails, a single membrane-spanning region and a large extracellular region containing three divalent-cation consensus sequences (Figure 1). The shorter and structurally different CD18 subunit has a characteristic protease-resistant cysteine-rich segment in the extracellular region in common with all integrin β subunits.[6]

Leukocyte functions mediated by β2 integrins (Table 1) include chemotaxis, phagocytosis, homotypic- and heterotypic adhesion, and antibody-dependent and independent cell-mediated cytotoxicity. β2 integrins bind to several ligands including iC3b, fibrinogen and factor X, to membrane receptors (e.g. CD54 and ICAM-2) and to bacterial and parasitic structures (e.g. lipid A, polysaccharides, Leishmenia gp 63)(reviewed in[1]).

Table 1. Leukocyte functions mediated by β2 integrins.

Myeloid cells	Lymphoid cells
Binding to iC3b & other ligands/substrates	Binding to ICAMs
Chemotaxis	Lymphocyte homing
Phagocytosis	Lymphoproliferation
Aggregation	Aggregation
Adhesion to endothelium	Adhesion to endothelium
Antibody-dependent cytotoxicity (ADCC)	Natural killing (NK) & cytotoxicity (CTL)

The biologic importance of β2 integrins in leukocytes was elucidated through the identification of an inherited deficiency of this receptor family in humans (Leu-CAM deficiency).[7] In this disease, structural abnormalities in the common CD18 subunit prevent heterodimer formation in the endoplasmic reticulum resulting in the inability of leukocytes to express all three heterodimers on the cell surface. This defect has devastating consequences to the function of phagocytic cells (polymorphonuclear leukocytes, monocytes, macrophages), evident despite the presence of persistent neutrophilia. The phagocytic cell defects are manifested clinically by recurrent and often fatal bacterial infections. Biopsies of infected tissues often reveal numerous bacteria, lymphocytes and plasma cells with very few neutrophils. Blood vessels at the inflammatory sites however are often congested, dilated and contain numerous neutrophils. Leu-CAM deficiency leaves the *in vivo* functions of lymphocytes relatively intact, as reflected by intact delayed hypersensitivity reactions, and clinically by the lack of increased susceptibility to viral infections. We have interpreted these findings as indicative of the presence of alternative adhesion pathways that can preserve lymphocyte functions *in vivo*[8,] an interpretation borne out by recent data (reviewed in[9]).

To subserve their vital role in chemotaxis, transmigration and phagocytosis, the low-avidity β2 integrin receptors on resting cells must rapidly and reversibly increase their avidity in response to environmental stimuli, rapidly redistribute on the surface membrane and tether to the force-generating contractile elements of the cytoskeleton during locomotion. These dynamic functions reflect the presence of complex regulatory pathways which modulate the activation state of β2 integrins, their surface distribution during cell migration and phagocytosis and their interaction with cytoskeleton. In addition to these functional adaptations of existing surface-membrane receptors, phagocytic cells have developed mechanisms for rapid recruitment of additional receptors from intracellular pools when needed. This latent pool of receptors is normally stored within 2° and 3° granules in neutrophils and is rapidly mobilized to the cell surface in a dose-responsive manner upon cell activation.[8, 10]

CONFORMATIONAL STATES ASSOCIATED WITH INCREASED RECEPTOR AVIDITY

As in other integrins,[11] increased β2 integrin avidity secondary to conformational changes in the extracellular domain can be induced in the purified receptors.[12] Thus mM concentrations of Mn^{++} and certain lipid extracts increase avidity of purified CD11b/CD18 to iC3b[12,13] and the avidity of CD11a/CD18 to CD54.[14] Agonists also induce expression of a Ca^{++}-dependent epitope for the NKI-L16 mAb on CD11a/CD18 in a temperature- and energy-independent manner and in the presence of inhibitors of mediators of intracellular signalling as PKC and PKA,[15] suggesting little role for additional proteins or post-translational modifications in rapid switching of integrins to a high avidity state and arguing that this state is an intrinsic feature of these receptors. While these conformational changes can be induced with certain mAbs or with Mn^{++}, induction of these states in the intact cell in the presence of physiologic plasma concentrations of cations, often requires metabolic energy. Thus expression of the 7E3 neoepitope on the high avidity state of CD11b/CD18 in the absence of Mn^{++} requires activation of cells with certain agonists and is dependent on both temperature and Ca^{++}- (but not Mg^{++}).[16] Expression of the activation epitope for mAb 24, common to all three CD11 subunits, also requires metabolic energy and mM concentrations of Mg^{++} and appears to be inhibited by Ca^{++}.[14] The structural basis for these altered conformational states is not understood. The requirements for metabolic energy argue however that intracellular signals can induce similar changes in these receptors under physiologic conditions.

ROLE OF THE CYTOPLASMIC TAIL OF CD18 IN SIGNALLING

Intracellular signal transduction involving G proteins, calcium, phospholipids and kinases have been implicated in the oscillations of integrins between low and high avidity states and appear to be mediated through modulation of the short cytoplasmic tails of these receptors.[17, 18, 19, 20, 21] Direct activation of protein kinase C through use of phorbol esters or indirectly, through triggering of several receptors as CD2 and CD3 in lymphocytes[22] or chemotactic receptors in neutrophils[18] switches β2 integrins from a low to a high avidity state.The high avidity state is short in duration, usually reversible in minutes to hours depending on the stimulus and is effectively blocked by staurosporin, a PKC inhibitor. Several observations pointed to the cytoplasmic tail of the common CD18 subunit as important in these rapid avidity changes. Activation of leukocytes with phorbol esters, or with the chemoattractant FMLP leads to rapid phosphorylation of the normally unphosphorylated CD18 on serine, threonine and tyrosine residues in its cytoplasmic tail[18, 23] (Figure 2).

This modification peaks within seconds and is reversible with kinetics similar to the changes in receptor avidity. Cell activation on the other hand does not modify the intrinsically phosphorylated state of the cytoplasmic tails of the associated CD11 subunits.

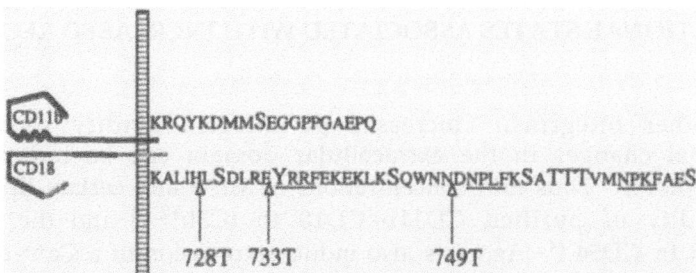

Figure 2. Schematic of the structure of CD11b/CD18 and the amino acid sequence of the cytoplasmic tails of its subunits. Caps outline potential phosphorylation sites. Arrows point to the positions of the introduced truncations in CD18 (see below). Putative internalization signals are underlined.

More recently, expression of the recombinant heterodimers in COS cells provided further support for a direct role of the cytoplasmic tail of CD18 in regulation of receptor avidity. The recombinant receptors expressed in these cells are intrinsically active when compared to the nascent receptors expressed in resting leukocytes.[20, 24] These receptors express activation epitopes (NKI-L16 for CD11a/CD18 and 7E3 for CD11b/CD18), are not further activated by phorbol esters and the cytoplasmic tail of the common CD18 subunit is intrinsically phosphorylated. Deletions in the cytoplasmic tail of CD18 had deleterious effects on ligand binding of CD11a/CD18 expressed in COS cells and abolished receptor activation by phorbol esters when the receptor was expressed in lymphocytes.[20] A similar analysis of CD11b/CD18 carried out in COS cells however revealed that receptor avidity evaluated by binding to iC3b varied depending on the site of CD18 truncation.[24] Removing approximately the C-terminal half of the cytoplasmic tail (749T, indicating a truncation introduced after residue 749 of the normal sequence, Figure 2) reduced binding by 2 fold, suggesting that the deleted segment is important in enhancing ligand binding of the exoplasmic domain of the WT receptors. Additional deletions however led first to a normalization (733T) and then to a 2 fold increase (728T) in ligand binding compared to WT receptors. The differences in the effects of truncations in the same CD18 subunit on ligand binding in CD11b/CD18 and CD11a/CD18 may reflect different regulatory pathways for members of the β2 integrins or critical roles for certain residues in regulating receptor avidity. A phosphorylated serine for example is included in the 731T construct of CD11a/CD18 which binds weakly to CD54.[20] This serine is deleted in 728T where binding to iC3b is increased. These data thus suggest the presence of specific regions of CD18 which can positively or negatively modulate receptor avidity perhaps by serving as recognition sites for cytoplasmic proteins. The preserved ligand binding ability of the mutant receptors observed in our study is also consistent with previous data in β1 integrins where similar truncations of the homologous cytoplasmic tail of the β1 subunit did not affect ligand binding.[19, 25] The increased binding of CD11b/CD18 observed in 728T with the shortest cytoplasmic tail (only five amino acids) may also explain our finding that removal of the whole cytoplasmic tail and transmembrane segments of both the CD11 and CD18 heterodimers still results in a functional receptor effective in ligand binding and in inhibiting neutrophil adhesion to cytokine-activated endothelium,[26] and is consistent with the idea that these receptors are normally kept in an inactive state through associations with one or more intracellular moieties. Release from this inhibitory effect could be induced by some of the truncations produced or by certain mAbs and divalent cations.

ROLE OF THE CYTOPLASMIC TAIL OF CD18 IN COATED PIT INTERNALIZATION

A characteristic feature of several receptors which undergo internalization through coated pits, is the presence of specific recognition sequences in their cytoplasmic tails for various adaptins. These sequences include but are not restricted to NPXY/F (in the LDL receptor, insulin- and some other growth factor receptors)[27] and YXRF (in the transferrin receptor).[28] The cytoplasmic tail of CD18, but not any of the CD11 subunits, contains two NPXF and one YXRF sequences of unknown function. We therefore tested if the recombinant WT receptor is internalized through coated pits and determined the effects of the various truncations described above on this process.[24] The WT receptor was immunostained with either an anti-CD11b or anti-CD18 mAbs at 4°C. Cells were then washed and warmed to 37°C for 10 minutes and one micron sections of each of 8-10 cells were examined by confocal microscopy. In cells immunostained and kept at 4°C, WT receptor was uniformly distributed. At 37°C, the receptor became clustered and approximately 40% was seen inside cells. In contrast, little or no endocytosis was seen in any of the three truncated forms described in the previous section. The lack of internalization of the 749T receptor which contains an intact YXRF motif similar to that found in the transferrin receptor[28] suggests that this motif is not functional in CD18. Conservation of the NPXF sequence in the β1 family suggests that internalization in this case may also occur through this same motif. Additional mutagenesis experiments will be required to fine map the relative contribution of each of the two NPXY/F motifs and surrounding amino acids. Other integrins in the β1 family are rapidly endocytosed through coated pits (t 1/2 of 7 min.) and recycle back to the plasma membrane at the leading edge of migrating fibroblasts.[29, 30, 31] Although it is premature to speculate about potential roles for endocytosis and/or recycling in β2 integrins, it is intriguing that endocytosis of β1 integrins is important in cell signalling.[32] Endocytosis and recycling may also play a role in selective membrane targeting of those receptors not linked to cytoskeleton or in their degradation. The role of receptor phosphorylation on such functions remains to be determined.

IDENTITY OF THE CYTOSKELETAL PROTEINS INTERACTING WITH CD18

Previous studies have identified several cytoskeletal proteins which colocalize with integrins at focal contact sites. These include talin, vinculin, α-actinin and actin.[33] In leukocytes, co-capping of CD11a/CD18 and talin occurs following phorbol ester treatment.[34] To begin to understand how integrin interactions with cytoskeleton are regulated, we attempted to determine the spectrum of cytoskeletal proteins which interact with the cytoplasmic tails of β2 integrins using an approach similar to the one used by Burridge and colleagues for β1.[33] The full cytoplasmic tails of human CD18 and of human CD11b were synthesized and covalently linked to CNBr-sepharose (CD18-sepharose and CD11b-sepharose, respectively). Glycine- or BSA sepharose were used as controls. Human monocyte detergent-soluble extracts were applied to these resins. After a 30-minute incubation at room temperature (RT) in the presence of protease- and phosphatase inhibitors, resins were washed extensively and electrophoresed on SDS PAGE. Major coomassie-stained protein bands at ~200 kD, 130-100 kD and 42 kD were visible in eluate from CD18 but not from CD11b- or control gels.[35] Western blot analysis identified these bands as ABP-280, talin, vinculin, α-actinin, gelsolin and actin respectively. None of these proteins were detected in eluates from CD11b- or control sepharose. It is likely that these proteins may interact with CD18 as a complex, or interact sequentially following agonist-induced redistribution of cytoskeletal elements. In either case, such interactions may facilitate receptor clustering, may help establish transmembrane connections between the ligated receptors and the force-generating cytoskeleton and may contribute to its organization.

Some or all of the cytoskeletal proteins detected in eluate from CD18-sepharose may bind directly or indirectly to this integrin. We have begun studies to evaluate the nature of the interaction of each of these proteins with CD18 and map the respective sites on the cytoplasmic tail. Preliminary data have shown that α-actinin binds directly to the cytoplasmic tail of CD18 as was previously shown with β1 integrins.[33] Purified α-actinin was also shown to bind to the intact CD11b/CD18 heterodimer immunoprecipitated from the detergent-soluble fraction of leukocytes, suggesting that this interaction may occur under more physiologic conditions . We have also mapped the α-actinin binding region in the cytoplasmic tail of CD18 to a 10 amino acid segment (residues 733-742).[35] Similar analysis of the interaction of the other cytoskeletal proteins with CD18 will be required before a comprehensive understanding of integrin association with cytoskeleton can be derived.

SUMMARY AND CONCLUSION

The complex functions played by β2 integrins in mediating a large variety of adhesive interactions of leukocytes are highly regulated. This regulation results in transient adaptations/associations, permitting physical and functional recycling of these receptors during chemotaxis, phagocytosis and target-cell killing. The structural definition of these adaptations will lead not only to a better understanding of how these receptors are regulated in leukocytes but also shed valuable light on how these integrins integrate diverse extracellular signals into spatially and temporaly coordinated cellular responses.

ACKNOWLEDGEMENTS

The authors acknowledge the secretarial assistance of Ms. Elena Fiamma and the support from the NIH, the March of Dimes and Birth Defects, Yamada Science Foundation of Japan and the American Heart Association, Massachusetts Affiliate.

REFERENCES

1. M.A. Arnaout, Structure and function of the leukocyte adhesion molecules CD11/CD18, *Blood* 75: 1037 (1990).
2. N. Kieffer and D.R. Phillips, Platelet membrane glycoproteins: Functions in cellular interactions, *Ann. Rev. Cell Biol.* 6: 329 (1990).
3. M.V. Nermut, N.M. Green, P. Eason, S.S. Yamada and K.M. Yamada, Electron microscopy and structural model of human fibronectin receptor, *EMBO J.* 7: 4093 (1988).
4. J.J. Calvete, A. Henschen and J. Gonzalez-Rodriguez, Assignment of disulphide bonds in human platelet GPIIIa: A disulphide pattern for the β-subunits of the integrin family, *Biochem. J.* 274: 63 (1991).
5. C.E. Nelson, H. Rabb and M.A. Arnaout, Genetic cause of leukocyte adhesion molecule deficiency: Abnormal splicing and a missense mutation in a conserved region of CD18 impair surface expression of β2 integrins.*J. Biol. Chem.* In press (1992).
6. R.O. Hynes, Integrins: a family of cell surface receptors, *Cell* 48: 549 (1987).
7. M.A. Arnaout, Leukocyte adhesion molecules deficiency: Its structural basis, pathophysiology and implications for modulating the inflammatory response, *Immunol. Rev.* 114: 145 (1990).
8. M.A. Arnaout, H. Spits, C. Terhorst, J. Pitt and R.F. Todd III, Deficiency of a leukocyte surface glycoprotein (LFA-1) in two patients with Mo1 deficiency: Effects of cell activation on Mo1/LFA-1 surface expression in normal and deficient leukocytes, *J. Clin. Invest.* 74: 1291 (1984).
9. L.M. Stoolman, Adhesion molecules controlling lymphocyte migration, *Cell* 56: 907 (1989).
10. R.F. Todd III, M.A. Arnaout, R.E. Rosin, C.A. Crowley, W.A. Peters, J.T. Curnuttee, *et al.*, Subcellular localization of the α subunit of Mo1 (Mo1 alpha; formerly gp110), a surface glycoprotein associated with neutrophil adhesion, *J. Clin. Invest.* 74: 1280 (1984).
11. P.J. Sims, M.H. Ginsberg, E.F. Plow and S.J. Shattil, Effect of platelet activation on the conformation of the plasma membrane glycoprotein IIb-IIIa complex, *J. Biol. Chem.* 266: 7345 (1991).
12. D.C. Altieri, Occupancy of CD11b/CD18 (Mac-1) divalent ion binding site(s) induces leukocyte adhesion, *J. Immunol.* 147: 1891 (1991).
13. A. Hermanowski-Vosatka, J.A.G. Van Strip, W.J. Swiggard and S.D. Wright, Integrin modulation factor-1: A lipid that alters the function of leukocyte integrins, *Cell* 68: 341 (1992).
14. I. Dransfield, C. Cabanas, A. Craig and N. Hogg, Divalent cation regulation of the function of the leukocyte integrin LFA-1, *J. Cell Biol.* 116: 219 (1992).
15. Y. van Kooyk, P. Weder, F. Hogervorst, A.J. Verhoeven, G. van Seventer, A.A. te Velde, *et al.*, Activation of LFA-1 through a Ca^{2+}dependent epitope stimulates lymphocyte adhesion, *J. Cell Biol.* 112: 345 (1991).
16. D.C. Altieri, R. Bader, P.M. Mannucci and T.S. Edgington, Oligospecificity of the cellular adhesion receptor MAC-1 encompasses an inducible recognition specificity for fibrinogen, *J. Cell Biol.* . 107: 1893 (1988).
17. J.W. Tamkun, D.W. DeSimone, D. Fonda, R.S. Pateb, C. Buck, A.R. Horwitz,

et al., Structure of integrin, a glycoprotein involved in the transmembrane linkage between fibronectin and actin, *Cell* 46: 271 (1986).

18. T. Chatila, R.S. Geha and M.A. Arnaout, Constitutive and stimulus-induced phosphorylation of CD11/CD18 leukocyte adhesion molecules, *J. Cell Biol.* 109: 3435 (1989).

19. Y. Hayashi, B. Haimovich, A. Reszka, D. Boettiger and A. Horwitz, Expression and function of chicken integrin β1 subunit and its cytoplasmic domain mutants in mouse NIH 3T3 cells, *J. Cell Biol.* 110: 175 (1990).

20. M.L. Hibbs, S. Jakes, S.A. Stacker, R.W. Wallace and T.A. Springer, The cytoplasmic domain of the integrin lymphocyte-function-associated antigen 1 β subunit: Sites required for binding to intercellular adhesion molecule 1 and the phorbol ester-stimulated phosphorylation site, *J. Exp. Med.* 174: 1227 (1991).

21. T.E. O'Toole, D. Mandelman, J. Forsyth, S.J. Shattil, E.F. Plow and M.H. Ginsberg, Modulation of the affinity of integrin αIIbβ3 (GPIIb-IIIa) by the cytoplasmic domain of αIIb, *Science* 254: 845 (1991).

22. Y. van Kooyk, P. van de Wiel-van Kemenade, P. Weder, T.W. Kuijpers and C.G. Figdor, Enhancement of LFA-1 mediated cell adhesion by triggering through CD2 or CD3 on T lymphocytes, *Nature* 342: 811 (1989).

23. J.P. Buyon, S.G. Slade, J. Reibman, S.B. Abramson, M.R. Philips, G. Weismann, *et al.*, Constitutive and induced phosphorylation of the α- and β-chains of the CD11/CD18 leukocyte integrin family: relationship to adhesion-dependent functions, *J. Immunol.* 144: 191 (1990).

24. H. Rabb, C.P. Sharma, M. Michishita, D. Brown and M.A. Arnaout, Regulation of CD11b/CD18 function and endocytosis by the cytoplasmic tail of its β subunit, *submitted* . (1992).

25. J. Solowska, J.-L. Guan, E.E. Marcantonio, J.E. Trevithick, C.A. Buck and R.O. Hynes, Expression of normal and mutant avian integrin subunits in rodent cells, *J. Cell Biol.* . 109: 853 (1989).

26. N. Dana, D.F. Fathallah and M.A. Arnaout, Expression of a soluble and functional form of the human β2 integrin CD11b/CD18, *Proc. Natl. Acad. Sci. (USA)* . 88: 3106 (1991).

27. W-J. Chen, J.L. Goldstein and M.S. Brown, NPXY, a sequence often found in cytoplasmic tails, is required for coated pit-mediated internalization of the low density lipoprotein receptor, *J. Biol. Chem.* 265: 3116 (1990).

28. J.F. Collawn, M. Stangel, L.A. Kuhn, V. Esekogwu, S. Jing, I.S. Trowbridge, J.A.Tainer., Transferrin receptor internalization sequence YXRF implicates a tight turn as a structural recognition motif for endocytosis, *Cell* 63: 1061 (1990).

29. M.S. Bretscher, Endocytosis and recycling of the fibronectin receptor in CHO cells, *EMBO J.* 8: 1341 (1989).

30. T.J. Raub, S.L. and Kuentzel, Kinetic and morphological evidence for endocytosis of mammalian cell integrin receptors by using an anti-fibronectin receptor β subunit monoclonal antibody, *Exp. Cell Res.* 184: 407 (1989).

31. M.M. Sczekan and R.L. Juliano, Internalization of the fibronectin receptor is a constitutive process, *J. Cell Physiol.* 142: 574 (1990).

32. M.M. Schwartz, C. Lechene and D.E. Ingbar, Insoluble fibronectin activates the Na/H antiporter by clustering and immobilizing integrin $\alpha5\beta1$, independent of cell shape, *Proc. Natl. Acad. Sci. (USA).* 88:7849 (1991).
33. C.A. Otey, F.M. Pavalko and K. Burridge, An interaction between α-actinin and the $\beta1$ integrin subunit in vitro, *J. Cell Biol.* 111: 721 (1990).
34. A. Kupfer and S.J. Singer, Molecular dynamics in the membranes of helper T cells, *Proc. Natl. Acad. Sci. (USA).* 85:8216 (1988).
35. C.P. Sharma, S. Magil and M.A. Arnaout, Microdomains in the cytoplasmic tail of CD18 involved in binding to cytoskeleton, *Clin Res.* . In press (1992).

LEUKOCYTE-ENDOTHELIAL CELL ADHESION AS AN ACTIVE, MULTI-STEP PROCESS: A COMBINATORIAL MECHANISM FOR SPECIFICITY AND DIVERSITY IN LEUKOCYTE TARGETING

Eugene C. Butcher

Laboratory of Immunology and Vascular Biology, Department of Pathology, Stanford University Medical Center, Stanford, CA 94305-5324; and the Center for Molecular Biology in Medicine, Veterans Administration Medical Center, Foothill Research Park (154B), Palo Alto, CA 94304

INTRODUCTION

The recruitment of leukocytes from the blood is one of the most dramatic cellular responses to tissue damage and inflammation, and is central to the physiologic trafficking of lymphocytes. Leukocyte extravasation is exquisitely regulated *in vivo* by mechanisms of selective leukocyte-endothelial cell (EC) recognition. Such selective recognition events help determine the degree and character of local inflammatory reactions by controlling the extravasation of neutrophils, monocytes and other leukocytes; facilitate immune surveillance by directing the recirculation pathways of virgin and memory lymphocytes; and support regional immune responses by targeting lymphocyte effector cells to particular organs and tissues in the body. The interaction of lymphocytes and leukocytes with the vascular endothelium *in vivo* can display extraordinary specificity in relation to the inflammatory stimulus, the stage of the inflammatory response, and the tissue site or organ involved. Examples include the almost exclusive attachment of eosinophils to venules in allergic reactions, the specific recruitment of monocytes to cerebral vessels in mouse models of cerebral malaria, and the tissue-selective interaction of lymphocyte subsets with high endothelial venules (HEV) in organized lymphoid tissues.

A number of adhesion receptors (ARs) have been identified that can participate in lymphocyte and/or leukocyte-EC interactions (reviewed in Butcher, 1990; Springer, 1990). Some of these are discussed below. Analyses of the *in vivo* expression and function of several of these AR's have confirmed their involvement in lymphocyte and leukocyte

Mechanisms of Lymphocyte Activation and Immune Regulation IV: Cellular Communications
Edited by S. Gupta and T.A. Waldmann, Plenum Press, New York, 1992

181

trafficking; however, apparent paradoxes have arisen that cannot be explained by simple lock-and-key models of leukocyte-EC recognition via adhesion molecules. For example, peripheral lymph node (PLN) HEV can bind both lymphocytes and neutrophils via the same molecular pathway (in which the leukocyte lectin homing receptor, the L-selectin (LECAM-1, LAM-1, the Mel-14 antigen), interacts with carbohydrate determinants of the PLN vascular addressin), yet only lymphocytes adhere stably to HEV and enter lymph nodes from the blood *in vivo* (Lewinsohn, 1987; Berg, 1991a,b). Similarly, the vascular E-selectin (ELAM-1) binds both skin-homing memory T cells and neutrophils, yet is thought to recruit neutrophils selectively during acute inflammation, and cutaneous memory T cells during chronic inflammation in the skin (Bevilacqua, 1989; Picker, 1991a; Shimizu, 1991). These observations seem to require the existence of additional levels of control during physiologic leukocyte-EC interactions.

Here we present a general model of leukocyte-endothelial cell recognition in which recognition is viewed as an active process requiring at least three sequential events (see Figure 1, and Butcher 1991). First, interaction is initiated by binding of constitutively functional leukocyte AR's to endothelial cell counter-receptors. In the best characterized examples, such primary adhesion is mediated by lectin-carbohydrate interactions involving leukocyte or vascular "selectins" and their cognate oligosaccharide ligands (see Figure 2). This initial adhesion is transient and reversible, unless followed by a second event--- activation of the leukocyte by specific chemoattractant or cell contact-mediated signals capable of triggering secondary AR's whose function is activation-dependent. Interaction of the activation-dependent AR with its EC counter receptor, the third step, results in strong, sustained attachment, completing the process of recognition. The best characterized activation-dependent ARs are heterodimeric integrins of the ß2 (CD18) or ß1 (CD29) classes. The model implies that leukocyte-EC recognition and extravasation can be controlled at any one of the three steps involved, and thus allows individual AR's to participate in multiple leukocyte homing events without sacrificing specificity. It also implies the possibility of generating a large number of specific recognition/homing events through combinatorial diversity from a limited complement of AR and chemoattractant/ receptor pairs.

Neutrophil Recognition of Inflamed Venules as a Paradigm

Each of the three steps - - - reversible adhesion, leukocyte activation, and activation-dependent binding - - - can be visualized during neutrophil interactions with venules in acute inflammation.

Step 1. Initial Interaction Consists of Reversible "Rolling" Mediated by the Neutrophil L-Selectin.
Interactions between circulating neutrophils and inflamed venules can be observed by video microscopy in *ex vivo* preparations of the rabbit mesentery. Although neutrophils do not normally interact with the venular endothelium (Fiebig, 1991), within minutes after tissue manipulation many neutrophils can be seen to interact loosely with the venular wall, "rolling" along the affected segments. This rolling is inhibited by antibody or Fab fragments to the neutrophil L-selectin (von Andrian, 1991), and by a soluble L-selectin-immunoglobulin chimeric protein (Ley, 1991). While the induced EC receptors involved in earliest rolling have not been characterized in *in situ* studies as yet, it seems likely that the vascular P-selectin (GMP140, PADGEM, CD62) and E-selectin

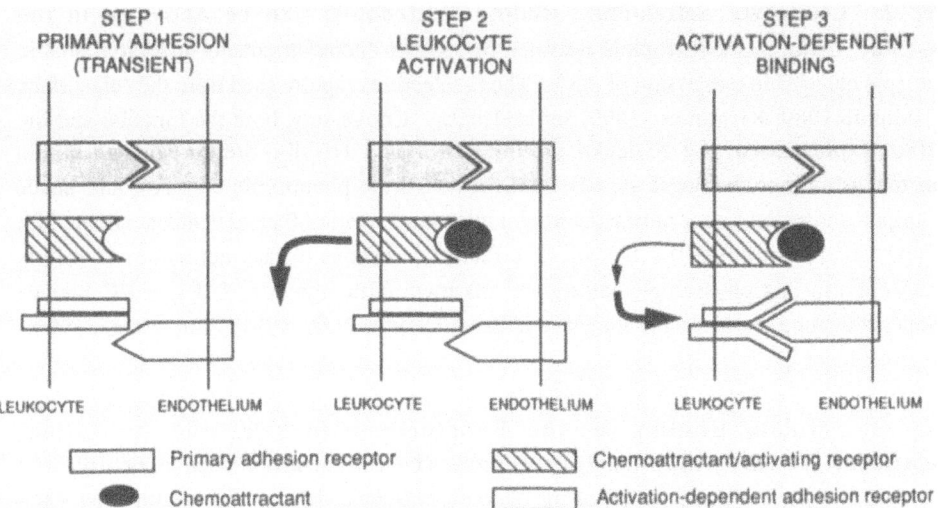

STEP 1
PRIMARY ADHESION
(TRANSIENT)

STEP 2
LEUKOCYTE
ACTIVATION

STEP 3
ACTIVATION-DEPENDENT
BINDING

LEUKOCYTE   ENDOTHELIUM   LEUKOCYTE   ENDOTHELIUM   LEUKOCYTE   ENDOTHELIUM

Primary adhesion receptor   Chemoattractant/activating receptor

Chemoattractant   Activation-dependent adhesion receptor

FIGURE 1. Schematic illustration of leukocyte-endothelial cell recognition as an active, three-step process. **Step 1.** Initial adhesion of blood borne leukocytes to venular endothelium involves interaction of constitutively functional leukocyte AR's. In the best characterized examples, such interactions involve lectins of the "selectin" family, interacting with oligosaccharide counter receptors (see Figure 2). This primary interaction, although unstable, occurs rapidly and can initiate leukocyte adhesion under flow conditions. It slows the transit of the leukocyte allowing sampling of the endothelium and local environment for activating signals. **Step 2.** If soluble or cell contact-mediated activating factors are present to which the leukocyte can respond, activation occurs leading to functional alterations in activation-dependent ARs. Although the figure depicts a soluble chemoattractant and its receptor, activating signals may also be delivered by endothelial cell surface molecules. **Step 3.** If the endothelium expresses counter-receptors for the triggered activation-dependent AR's, firm binding occurs that is stable under physiologic shear forces, completing the process of recognition. The best characterized activation-dependent AR's are the heterodimeric integrins of the ß1 (CD29) and ß2 (LeuCAM or CD18) classes. Their endothelial counter-receptors are often immunoglobulin-related, although these integrins can also mediate interactions with diverse extracellular matrix elements.

(ELAM-1) participate (Bevilacqua, 1989; Geng, 1990). These vascular AR's, which belong to the same C-type lectin/"selectin" gene family as the L-selectin, bind carbohydrate determinants of the neutrophil cell surface, including sialyl Lewis X (Walz, 1990; Brandley, 1990; Polley, 1991; Berg 1991d). Interestingly, recent *in vitro* studies suggest that one function of the L-selectin in neutrophil interactions with EC is to present these neutrophil carbohydrate ligands to the vascular selectins, a role for which it may be particularly well suited by virtue of its high degree of glycosylation and its unique topographic localization to microvillous sites of initial cellular contact (Picker, 1991b). Importantly, neutrophil L-selectin is constitutively functional (Lewinsohn, 1987). It is present at high levels on circulating, resting neutrophils, and mediates their attachment to cytokine-stimulated EC under flow conditions and in the absence of neutrophil activation (Hallmann, 1991; Smith, C.W., 1991).

Rolling greatly slows the transit of neutrophils through inflamed venules, allowing time for sampling of the local environment or the EC surface for activating/chemoattractant signals.

Step 2. Leukocyte Activation: Rolling Neutrophils Can be Activated in the Vascular Lumen. Neutrophils activated by specific chemoattractants undergo dramatic and rapid changes in expression of AR's. The L-selectin is rapidly shed from the cell surface (Kishimoto 1989; Kishimoto, 1990; Jutila, 1989a). Conversely, both the function and the surface expression of the ß2 leukocyte integrin Mac-1 ($\alpha_M\beta_2$) are upregulated within minutes. Immunohistologic studies confirm that these phenotypic changes, and hence neutrophil activation, occur during neutrophil interactions with inflamed venular endothelium *in vivo* (Kishimoto, 1989). The specific factor(s) responsible for activation of rolling neutrophils in this physiologic setting are unknown, although IL8/NAP1 or EC surface platelet activating factor are prominent candidates (Smith, W.B., 1991; Zimmerman, 1990 - see also Figure 2).

Step 3. Stable Binding to the Endothelium Involves the Activation-Dependent ß2 Integrin, Mac-1. As indicated above, neutrophil activation triggers a dramatic increase in surface levels of the ß2 integrin Mac-1. Mac-1, along with other AR's and enzymes, is stored within specialized granules of resting neutrophils, and is translocated upon activation to the cell surface by granule fusion with the plasmid membrane (Bainton, 1987). Neutrophil activation also triggers functional activation of preexisting cell surface Mac-1, presumably through conformational and/or topographical alterations (Buyon, 1988;

Step 1 Primary adhesion receptors		Step 2 Chemoattractant/activating factors	Step3 Activation-dependent adhesion receptors	
Leukocyte	**Endothelium**		**Leukocyte**	**Endothelium**
Lectin:carbohydrate		*Intercrine family*	*Integrins*	
L-selectin (L,N,M)	Lymph node addressin	Interleukin-8 (N,L)	LFA-1 ($\alpha L\beta 2$) (L>N,M)	ICAM-1,-2
CLA (smTL)	E-selectin (ELAM-1)	hMGSA/GROα (N)	Mac-1 ($\alpha M\beta 2$) (N,M,sL)	ICAM-1, others
sLe^x (N,M)	"	Platelet factor 4 (N,M)	p150,95 ($\alpha X\beta 2$) (N,M,sL)	?
"	P-selectin (GMP140,	RANTES (mTL,M)		
	CD62)	HuMIP–1α (CD8+ TL, BL)	VLA-4 ($\alpha 4\beta 1$) (M, mL>vL)	VCAM-1
Other		HuMIP–1β (vTL, ?M)		
?	Mucosal addressin	I-309 (M)		
		Monocyte chemoattractant protein-1 (M)		
		others		
		Lipids		
		Platelet activating factor (N)		
		Leukotriene B4 (N, M)		
		others		
		Other chemoattractants		
		C5a (N)		
		formyl peptides (N,M)		
		Interleukin-2 (sTL)		
		Cell contact-mediated		
		E-selectin binding (N)		
		CD44 (L)		

Figure 2. This Figure lists some adhesion receptor/ligand pairs and activating factors that are known to be able to or are likely to participate in leukocyte-endothelial cell interactions *in vivo*. This is not intended to be a comprehensive list. CD44, LPAM-1 (α4ßp) and HEBFpp are not included under adhesion receptors, for example, because their role in primary vs. activation-dependent adhesion events remains to be explored. For detailed descriptions of the ARs see text and reviews by Berg (1991) and Springer (1990); for a review of the chemoattractants see Schall (1991) and Oppenheim et al (1991). The leukocyte subset(s) that express each AR, or that have been shown to respond to each activating factor, are indicated in parentheses. N=neutrophils. M=monocytes. L=lymphocytes. BL=B lymphocytes. vTL and mTL=virgin and memory T lymphocytes. sTL indicates that only a subset of T cells responds.

Vedder, 1988). Indeed, in the absence of activating signals, Mac-1 and the other leukocyte ß$_2$ integrins are thought to be largely inactive (see for example Lawrence, 1991). Consistent with this, monoclonal antibodies (MAb's) to the leukocyte integrin ß2 chain have no effect on primary neutrophil rolling in inflamed venules. However, they completely prevent the subsequent arrest and stable attachment of neutrophils that precedes diapedesis. In the presence of anti-ß2 antibodies, neutrophils merely roll along inflamed endothelium, and are released to return to the circulation (Arfors, 1987; von Andrian, 1991). Thus *in situ* studies confirm the involvement of reversible primary adhesion, leukocyte activation, and secondary activation-triggered attachment mechanisms in neutrophil-EC recognition.

The requirement for each step in the recognition process has been elegantly modeled in *in vitro* studies of neutrophil interactions with activated cultured endothelium (Lawrence, 1990; Smith, C.W, 1991), and more recently with purified endothelial cell counter-receptors (Lawrence, 1991). Under flow conditions neutrophils adhere reversibly and "roll" on glass surfaces coated with P-selectin, a primary adhesion ligand. In contrast, under similar conditions neutrophils are unable to interact detectably with surfaces bearing ICAM-1, a ligand for the activation-dependent ß2 integrins, even if the flowing neutrophils are activated. The entire recognition cascade can be modeled, however, on slides coated with both P-selectin and ICAM-1: activation of rolling neutrophils with a phorbol ester results in their rapid arrest and firm adhesion through integrin binding to ICAM-1. These findings emphasize the essential and sequential roles of primary adhesion, activation, and activation-dependent attachment in the recognition process.

Can the Model be Generalized to Lymphocyte and Other Leukocyte EC Interactions?

Several lines of evidence suggest that the model is applicable not only to neutrophils, but also to monocytes, lymphocytes, and other leukocytes. Monocytes, like neutrophils, express primary AR's including the lectin L-selectin and carbohydrate ligands for the vascular selectins, and activation-dependent secondary AR's of both the ß1 and ß2 integrin classes, including Mac-1. Moreover, monocyte recruitment from the blood into the inflamed peritoneum, like neutrophil extravasation, is inhibited by anti-L-selectin and by anti-Mac-1 antibodies (Jutila, 1989b). This striking parallel with neutrophils clearly argues that monocyte extravasation also involves both activation-independent and activation-dependent interactions. Specificity of neutrophil vs. monocyte recruitment may depend to a large extent on specific activating signals: neutrophils and monocytes respond differentially to a number of known chemoattractants (see for example Figure 2). Monocytes also express higher levels of α4ß1 (VLA-4) than neutrophils, and thus may be able to bind preferentially to endothelial cells via the inducible EC ligand VCAM-1 (Elices, 1990).

Lymphocytes also express both primary and activation-dependent AR's, and employ them in interactions with EC in various settings: 1) Lymphocyte binding to HEV in peripheral lymph nodes, and lymphocyte extravasation or homing into lymph nodes *in vivo*, appears to be directed primarily by the L-selectin and its HEV ligand, the peripheral lymph node vascular addressin (PNAd) (reviewed in Berg, 1991a), but the activation-dependent integrin LFA-1 can also participate (Hamann, 1988). 2) Homing of lymphocytes via HEV in intestinal Peyer's patches involves the mucosal vascular addressin (MAd), which appears to be a primary adhesion receptor (Berg, 1991); and LFA-1 as well (Hamann, 1988). [α4 integrins also participate in HEV recognition in Peyer's patches (Holzmann, 1989), but it is unknown whether the role of the α4 integrin in this context is activation-dependent or independent (see Caveats and Comments, below).] Importantly, *in situ* observations of lymphocyte interactions with Peyer's patch HEV, visualized by intravital video microscopy, are consistent with the existence of reversible primary and stabilizing secondary adhesion events (Bjerknes, 1986). Initial lymphocyte adhesion to HEV is transient, generally lasting less than 1-2 seconds. Lymphocytes that fall off usually bind again downstream, seeming to skip along the endothelium. Only about 1 in 5 adherence proceeds to stable attachment. 3) Lymphocyte binding to cytokine-stimulated EC also involves activation-independent and activation-dependent mechanisms. Interaction of skin-homing memory T cells with interleukin-1-stimulated EC, for example, involves primary carbohydrate:E-selectin and activation-triggered LFA-1 and VLA4 integrin pathways (Picker, 1991; Shimizu, 1991). Antibodies to α4 integrins also inhibit the extravasation of memory lymphocytes into some sites of chronic inflammation *in vivo* (Issekutz, 1991), possibly by interfering with activation-linked VLA4 adhesion to induced EC VCAM-1 (Shimizu, 1991).

It is interesting that lymphocytes and monocytes express both ß2 and α4ß1 integrins, whereas neutrophils are ß2^{hi} but bear very little α4. This apparent redundancy in activation-dependent adhesion receptors on mononuclear cells may help explain the observation that, in patients with a genetic deficiency in ß2 expression, lymphocyte and monocyte traffic is qualitatively normal whereas neutrophil recruitment to inflammatory sites is severely depressed (Anderson, 1987). Lymphocyte homing may also take advantage of the fact that, although constitutively active, some primary AR's can be triggered to a higher affinity or avidity state following lymphocyte activation. This appears to be the case for the L-selectin lymph node homing receptor, for example (Spertini, 1991). Thus, at least in HEV where vascular addressins are expressed at very high levels, the L-selectin might be able to participate both in initial (reversible) adhesion and in activation-enhanced attachment.

Finally, a number of cell surface and secreted activating factors exist that act on distinct subsets of mononuclear cells, including lymphocytes, and that therefore may contribute to the selectivity of step two in lymphocyte-EC recognition (Figure 2). Signaling via engagement or crosslinking of some lymphocyte adhesion receptors could be important in this context (*e.g.* CD44---Koopman, 1990) . Of particular interest is the emerging "intercrine" superfamily of chemoattractants (reviewed in Schall, 1991; Oppenheim, 1991). The intercrine RANTES is a chemoattractant for memory but not naive (virgin) T cells. In contrast, the related HuMIP-1ß attracts naive T lymphocytes preferentially, whereas HuMIP-1α attracts CD8$^+$ T cells and B cells. Other intercrine members attract neutrophils preferentially, and several are chemoattractant for monocytes (see Figure 2). Many remain to be tested on lymphocyte subsets. Interestingly, many of the intercrines bind heparin, leading to the suggestion that they may be presented to leukocytes either as soluble chemoattractants or as immobilized molecules bound to glycosaminoglycans of the EC surface.

Although more direct tests of the model will be required in each case, these considerations suggest that reversible adhesion, activation, and activation-dependent binding mechanisms may prove to be common features of many or all leukocyte-endothelial cell recognition events.

Implications of the Model

Combinatorial Diversity in Leukocyte-EC Recognition. The model requires that three independent receptor/ligand pairs be occupied, sequentially but during the same encounter, in order for leukocyte recognition of EC to be complete. This requirement implies that the specificity of leukocyte-EC recognition is determined by unique combinations of primary adhesion, activating, and secondary adhesion receptor/ligand pairs. It follows that the number of possible specific leukocyte-EC recognition events is given by the product of the number of possible receptor/ligand pairs at each step. As listed in Figure 2, at present we know of at least 4-5 primary leukocyte-EC AR pairs, 12 defined lymphocyte or leukocyte chemoattractants, and 4 secondary, activation-dependent AR pairs. Certainly more remain to be discovered. If each of these known receptor/ligand pairs were expressed on unique subsets of leukocytes and EC, $5 \times 12 \times 4 = 240$ independent leukocyte-EC recognition events would be possible. Although such highly restricted expression of receptors is not common *in vivo* (see below), this calculation emphasizes the implications of a multi-step mechanism for combinatorial diversity in cellular recognition.

Successful Receptor/Ligand Pairs Can be Used in Multiple Specific Leukocyte-EC Recognition Events. The multi-step model of recognition permits the involvement of a given AR in more than one leukocyte-EC (or other cell-cell) recognition event, without sacrificing specificity. Indeed, this ability may be one of the most important advantages of the multi-step recognition process, given that the individual receptor/ligand pairs listed in Figure 2 are rarely (perhaps never) restricted in usage to a single physiologic leukocyte-EC recognition event. Examples of involvement of primary adhesion pathways in 2 or more distinct leukocyte trafficking events were mentioned above, and indeed helped to force the formalization of the multi-step model presented here. The expression of the activation-dependent AR's and their ligands is even more widespread. In fact, the leukocyte integrins serve a wide variety of leukocyte cell-cell and cell-substrate adhesion events (reviewed in Arnaout, 1990; Hemler, 1990; Shimizu, 1990; Springer, 1990).

The Specificity of the Recognition Process Can Exceed That of its Component Steps. In its purest form, the model would permit absolute specificity of leukocyte and lymphocyte homing through all-or-none regulation of the receptors involved. In reality, however, there appears to be considerable "play" in the system. The expression of AR's on leukocyte subsets and on EC in different vascular beds can be quite specific, but just as often differs more quantitatively than qualitatively. Thus the combinatorial determination of leukocyte-EC recognition specificities operates more in an analogue than in a binary digital world. Such "play" may reflect an evolutionary balance between the need for specificity and, at the same time, for redundancy in cell-cell recognition mechanisms. In this context, one of the predicted strengths of the multi-step recognition process is that the specificity of leukocyte-EC recognition and extravasation can exceed the specificity of any of its components. For example, if in a given inflamed vessel leukocyte A had four times the probability of interacting effectively at all 3 steps compared to leukocyte B, then leukocyte A would bind stably 64 times more often. The model also predicts that recognition is inherently resilient. If for example selectivity for leukocyte A vs. B were lost at one step through an evolutionary quirk (i.e., if leukocyte B now expressed a receptor previously restricted to leukocyte A), leukocyte A would still bind 16 times as well as leukocyte B.

New Homing Specificities Can be Derived From Novel Combinations of Existing AR and Chemoattractant/Receptor Pairs.   Another implication of the combinatorial determination of specificity is that novel leukocyte-EC recognition events can be generated during evolution without the necessity of evolving new adhesion receptors. Instead, new recognition specificities could be selected merely by expressing existing receptor pairs in novel combinations. The evolution or selection of new specificities would be facilitated in situations where the receptors involved in one or more steps are fairly broadly expressed.   For example, if all memory T cell homing shared common chemoattractant and secondary AR mechanisms, then the generation of new subset homing patterns would only require altered regulation of primary adhesion pathways.

Possible Additional Steps in the Regulation of EC Recognition and Extravasation

Intracellular Signal Transduction Mechanisms Could Provide an Additional Regulated Step Linking Leukocyte Activation to Individual Activation-Dependent ARs.   The model as proposed assumes that leukocyte activation triggers all activation-dependent surface adhesion mechanisms simultaneously. However, more specific signal transduction linkages between chemoattractant receptors and activation-dependent AR's may be possible, allowing independent activation of particular adhesion mechanisms by different activating stimuli. If such mechanisms do exist they would constitute a fourth step in the process, allowing even further combinatorial diversity in leukocyte-EC recognition.

Events Following Endothelial Cell Recognition May Regulate Leukocyte Recruitment, As Well.   Activation-dependent adhesion via leukocyte integrins, although stable under physiologic shear forces and sustained for minutes, is eventually reversible (reviewed in Arnaout, 1990; Shimizu, 1990). For example, activation-triggered Mac-1-mediated adhesion of neutrophils to cultured EC decays spontaneously, with release of bound cells after 10-15 minutes (Lo, 1989). This reversibility suggests that extravasation need not obligatorily follow EC recognition, although these events could be linked in some settings---for example if a common chemoattractant, arriving in the vessel from an extravascular source, were the signal both for activation-dependent adhesion to EC and for diapedesis and migration into the tissues.   If instead the activating signal of step 2 in EC recognition were provided by the endothelium itself, then independent tissue-derived chemoattractant gradients would likely be required for extravasation. Thus in some settings extravascular signals may be able to determine whether a leukocyte bound to the vascular endothelium eventually extravasates, or instead is freed to return to the circulation, thus adding yet another level to the multi-step control of leukocyte recruitment from the blood.

Caveats and Comments

In its simplest form, the model proposes that primary adhesion is necessary to slow the passage of blood-borne leukocytes so that they can sample the local microenvironment (or the EC surface) for activating signals, and perhaps so that activation-dependent AR's can engage their ligands. The requirement for a specific activation-independent adhesion event would be eliminated, however, in the setting of reduced blood flow, for example in venous stasis, thrombosis, or shock. Consistent with this concept, activation-dependent mechanisms are sufficient to initiate leukocyte binding to stimulated EC under static conditions *in vitro*.

In the simplest form of the model, the receptor/ligand pairs involved in each step of the recognition process are regarded to be distinct. In many instances, however, molecules may be able to subserve more than one role in the process. For example, engagement or crosslinking of primary adhesion receptors on the leukocyte could itself deliver an integrin-activating signal, as proposed for activation of neutrophils by the vascular E-selectin (Lo, 1991). Furthermore, certain adhesion receptors can apparently alter their affinity, and even their ligand specificity, upon activation of the cells expressing them. As mentioned above, the L-selectin, for example, may be triggered to a higher affinity or avidity state following lymphocyte activation (Spertini, 1991), and thus may be able to participate both in initial (reversible) adhesion and in activation-enhanced attachment. GPIIb-IIIa, a platelet integrin, binds immobilized fibronectin constitutively, but displays induced adhesiveness for a number of other ligands upon platelet activation (Phillips *et al.*, 1991). Thus certain integrins may prove to participate in leukocyte-endothelial cell recognition both as primary and as activation-dependent adhesion receptors, with overlapping or distinct EC ligands involved at each step.

Finally, it should be emphasized that, although we have limited our discussion here to events following initial leukocyte encounter with the vascular endothelium, the regulation of leukocyte extravasation and recruitment *in vivo* begins earlier. The expression of AR's on leukocyte subsets is determined in part developmentally and (in the case of lymphocytes) as a function of their prior history of antigen stimulation (reviewed in Butcher, 1986). The phenotype of the endothelium, and the array of local leukocyte activating factors that control the interactions of circulating leukocytes must necessarily be determined in advance as a function of the local tissue microenvironment, and of inflammatory stimuli. The role of inflammatory cytokines in regulating EC expression of AR's is well documented, for example (reviewed in Pober, 1990). Inflammatory signals also regulate the induction and expression of chemoattractant/activating factors in EC, and in surrounding tissue elements as well (Pober, 1990; Schall, 1991; Oppenheim, 1991).

CONCLUDING REMARKS

In conclusion, we have presented a model of leukocyte-endothelial cell recognition as an active process involving 3 or more sequential events leading from the initial encounter to stable attachment (Butcher 1991). It is likely that many variations on this theme exist, only some of which have been considered. The most important implications of the model, however, derive not so much from the specifics of the individual steps involved, but rather from the concept that recognition and stable attachment require several sequential events, each representing a go/no go decision point. It is the involvement of several steps, with multiple alternative receptor/ligand pairs available for participation at each one, that provides a combinatorial mechanism for generating both diversity and specificity in leukocyte-endothelial cell recognition; that permits molecular conservatism by allowing individual receptor/ligand pairs to participate in more than one independently regulated leukocyte homing event; and that implies that new homing specificities can be generated during evolution through the use of of existing receptor/ligand pairs in novel combinations.

ACKNOWLEDGMENTS

I thank the past and present members of my laboratory for their scientific and editorial input. Supported in part by NIH grants GM37734, AI19957, GM41965, and DK38707, UC Berkeley Tobacco-related Disease Research grant RT454, awards from the Veterans Administration and from Smith-Kline/Beecham, and an Established Investigator Award from the American Heart Association. This review was adapted from a previous manuscript, *Cell*, 67, 1033-1036, 1991.

References

Anderson, D.C. and Springer, T.A. Leukocyte adhesion deficiency: an inherited defect in the Mac-1, LFA-1, and p150,95 glycoprotein. *Annu. Rev. Med.* 3 8:175-194, 1987.

Arfors, K.E., Lundberg, C., Lindborm, L., Lundberg, K., Beatty, P.G., Harlan, J.M. A monoclonal antibody to the membrane glycoprotein complex CD18 inhibits polymor-phonuclear leukocyte accumulation and plasma leakage *in vivo*. *Blood,* 6 9:338-340, 1987.

Arnaout, M.A. Structure and function of the leukocyte adhesion molecules CD11/CD18. *Blood* 7 5:1037-1050, 1990.

Bainton, D.F., Miller, L.J., Kishimoto, T.K., Springer, T.A. Leukocyte adhesion receptors are stored in peroxidase-negative granules of human neutrophils. *J. Exp. Med.* 1987, **16** 6:1641-53, 1987.

Berg, E.L., Picker, L.J., Robinson, M.K., Streeter, P.R., Butcher, E.C. Vascular Addressins: Tissue selective endothelial cell adhesion molecules for lymphocyte homing. *Cellular and Molecular Mechanisms of Inflammation.* Vol. 2. Vascular Adhesion Molecules (C. Cochrane, M.A. Gimbrone Jr., eds.), Academic Press, Inc., San Diego, pp. 111-129, 1991.

Berg, E.L., Robinson, M.K., Warnock, R.A., and Butcher, E.C. The human peripheral lymph node vascular addressin is a ligand for LECAM-1, the peripheral lymph node homing receptor. *J. Cell. Biol.* **114**:343-349, 1991b.

Berg, E.L., Yoshino, T., Rott, L.S., Robinson, M.K., Warnock, R.A., Kishimoto, T.K., Picker, L.J., and Butcher, E.C. The cutaneous lymphocyte antigen is a skin lymphocyte homing receptor for the vascular lectin endothelial cell-leukocyte adhesion molecule 1. *J. Exp. Med.* **174**:1461-1466, 1991c.

Berg, E.L., Robinson, M.K., Mansson, O., Butcher, E.C., and Magnani, J.L. A carbohydrate domain common to both sialyl Le^a and sialyl Le^x is recognized by the endothelial cell leukocyte adhesion molecule, ELAM-1. *J. Biol. Chem.* **266**:14869-14872, 1991c.

Bevilacqua, M.P., Stengelin, S., Gimbrone, Jr., M.A., Seed, B. Endothelial leukocyte adhesion molecule 1: An inducible receptor for neutrophils related to complement regulatory proteins and lectins. *Science* **243**:1160-1165, 1989.

Bjerknes M; Cheng H; Ottaway C. Dynamics of lymphocyte-endothelial interactions *in vivo*. *Science*, **23**:402-405, 1986.

Brandley, B.K., Sweidler, S.J., Robbins Carbohydrate ligands of the LEC cell adhesion molecules. Cell **63**: 861-863.

Butcher, E.C. The regulation of lymphocyte traffic. In: Current Topics in Microbiology and Immunology **128**:85-122, 1986.

Butcher, E.C. Cellular and molecular mechanisms that direct leukocyte traffic. *Am. J. Pathology* **136**:3-11, 1990.

Butcher, E.C. Leukocyte-endothelial cell recognition: Three (or more) steps to specificity and diversity. *Cell* **67**:1033-1036, 1991.

Buyon, J.P., Abramson, S.B., Philips, M.R., Slade, S.G., Ross, G.D., Weissman, G., and Winchester, R.J. Dissociation between increased surface expression of gp165/95 and homotypic neutrophil aggregation. *J. Immunol.* **140**:3156-3160, 1988.

Elices M.J., Osborn, L., Takada, Y., Crouse, C., Luhowskyj, S., Hemler, M.E., Lobb, R.R. VCAM-1 on activated endothelium interacts with the leukocyte integrin VLA-4 at a site distinct from the VLA-4/fibronectin binding site. *Cell*, **60**:577-84, 1990.

Fiebig, E., Ley, K., and Arfors, K.-E. Rapid leukocyte accumulation by "spontaneous" rolling and adhesion in the exteriorized rabbit mesentery. *Int. J. Microcirc: Clin. Exp.* **10**:127-144, 1991.

Geng, J.G., Bevilacqua, M.P., Moore, K.L., McIntyre, T.M., Prescott, S.M., Kim, J.M., Bliss, G.A., Zimmerman, G.A., McEver, R.P. Rapid neutrophil adhesion to activated endothelium mediated by GMP-140. *Nature* **343**:757-60, 1990.

Hallmann, R., Jutila, M.A., Smith, C.W., Anderson, D.C., Kishimoto, T.K., and Butcher E.C. The peripheral lymph node homing receptor, LECAM-1, is involved in CD18-independent adhesion of human neutrophils to endothelium. *Biochem. & Biophysical Res. Comm.* **174**:236-243, 1991.

Hamann, A., Jablonski-Westrich, D., Duijvestijn, A., Butcher, E.C., Baisch, H., Harder, R., and Thiele, H.-G. Evidence for an accessory role of LFA-1 in lymphocyte-high endothelium interaction during homing. *J. Immunol.* **140**:693-699, 1988.

Hemler, M.E. VLA proteins in the integrin family: structures, functions, and their role on leukocytes. *Ann. Rev Immunol*, **8**:365-400, 1990.

Holzmann, B., McIntyre, B.W., Weissman, I.L. Identification of a murine Peyer's patch-specific lymphocyte homing receptor as an integrin molecule with an alpha chain homologous to human VLA-4 alpha. *Cell* **56**:37-46, 1989.

Imai, Y., Singer, M.S., Fennie, C., Lasky, L.A., and Rosen, S.D. Identification of a carbohydrate-based ligand for a lymphocyte homing receptor. *J. Cell Biol.* **113**:1213-1221, 1991.

Issekutz, T.B. Effects of anti-VLA-4 on lymphocyte migration by T cells with diverse homing properties to cutaneous and joint inflammation and to lymphoid tissues. *FASEB Journal* **5**: A1335, 1991.

Jutila, M.A., Rott, L., Berg, E.L., and Butcher, E.C. Function and regulation of the neutrophil MEL-14 antigen *in vivo*: comparison with LFA-1 and MAC-1. *J. Immunol.* **143**: 3318-3324, 1989a.

Jutila, M.A., Berg, E. L., Kishimoto, T.K., Picker, L.J., Bargatze, R.F., Bishop, D.K., Orosz, C.G., Wu, N.W., and Butcher, E.C. Inflammation induced endothelial cell adhesion to lymphocytes, neutrophils and monocytes: Role of homing receptors and other adhesion molecules. *Transplantation* **48**:727-731, 1989b.

Kishimoto, T.K., Jutila, M.A., and Butcher, E.C. Identification of a human peripheral lymph node homing receptor: A rapidly down-regulated adhesion molecule. *PNAS, USA* **87**:2244-2248, 1990.

Kishimoto, T.K., Jutila, M.A., Berg, E.L., and Butcher, E.C. Neutrophil Mac-1 and MEL-14 adhesion proteins inversely regulated by chemotactic factors. *Science* **245**:1238-1241, 1989.

Lawrence, M.B., Smith, C.W., Eskin, S.G., McIntire, L.V. Effect of venous shear stress on CD18-mediated neutrophil adhesion to cultured endothelium. *Blood* **75**:227-37, 1990.

Lawrence, M.B., Springer, T.A. Leukocytes roll on selectin at physiologic flow rates: Distinction from and prerequisite for adhesion through integrins. *Cell.*, **65**: 859-873, 1991.

Ley, K., Gaethgens, P., Fennie, C., Singer, M.S., Lasky, L.A., Rosen, S.D. Lectin-like cell adhesion molecule 1 mediates leukocyte rolling in mesenteric venules *in vivo. Blood* **77**:2553-2555, 1991.

Lewinsohn, D., Bargatze, R. and Butcher, E.C. Leukocyte-endothelial cell recognition: Evidence of a common molecular mechanism shared by neutrophils, lymphocytes, and other leukocytes. *J. Immunol.* **138**:4313-4321, 1987.

Ley, K., Gaethgens, P., Fennie, C., Singer, M.S., Lasky, L.A., Rosen, S.D. Lectin-like cell adhesion molecule 1 mediates leukocyte rolling in mesenteric venules *in vivo. Blood* **77**: 2553-2555, 1991a.

Lo, S.K., Detmers, P.A., Levin, S.M., and Wright, S.D. Transient adhesion of neutrophils to endothelium. *J. Exp. Med.* **169**:1779-1793, 1989.

Lo, S.K., Lee, S., Ramos, R.A., Lobb, R., Rosa, M., Chi-Rosso, G., Wright, S.D. Endothelial-leukocyte adhesion molecule 1 stimulates the adhesive activity of leukocyte integrin CR3 (CD11b/CD18, Mac-1, αMß2) on human neutrophils. *J. Exp. Med.* **173**: 1493-1500, 1991.

Oppenheim, J.J., Zachariae, C.O.C., Mukaida, N., and Matsushima, K. Properties of the novel proinflammatory supergene "intercrine" cytokine family. *Ann. Rev. Immunol.* **9**:617-648, 1991.

Phillips, D.R., Charo, I.F., and Scarborough, R.M. GPIIb-IIIa: The responsive integrin. *Cell* **65**:359-362, 1991.

Picker, L.J., Kishimoto, T.K., Smith, C.W., Warnock, R. A., and Butcher, E.C. ELAM-1 is an adhesion molecule for skin-homing T cells. *Nature* **349**:796, 1991a.

Picker, L.J., Warnock, R.A., Burns, A.R., Doerschuk, C.M., Berg, E.L., Butcher, E.C. The neutrophil selectin LECAM-1 presents carbohydrate ligands to the vascular selectins ELAM-1 and GMP-140. *Cell* **66**:921-933, 1991b.

Pober, J.S. and Cotran, R.S. The role of endothelial cells in inflammation. *Transplantation* **50**:537-544, 1990.

Polley, M.J., Phillips, M.L., Wayner, E., Nudelman, E., Singhal, A.K., Hakamori, S.I., and Paulson, J.C. CD62 and ELAM-1 recognize the same carbohydrate ligand, sialyl-Lex. *PNAS* **88**:6224-6228, 1991.

Schall, T.J. Biology of the RANTES/SIS cytokine family. *Cytokine* **3**:1-18, 1991.

Shimizu, Y., Shaw, S., Graber, N., Gopal, T.V., Horgan, K.J., Van Seventer, G.A., Newman, W. Activation-independent binding of human memory T cells to adhesion molecule ELAM-1 *Nature*, **349**:799-802, 1991

Shimizu, Y., Newman, W., Gopal, T.V., Horgan, K.J., Graber, N., Beall, L.D., van Seventer, G.A., Shaw, S. Four molecular pathways of T cell adhesion to endothelial cells: Roles of LFA-1, VCAM-1, and ELAM-1 and changes in pathway hierarchy under different activation conditions. *J. Cell Biol.* **113**:1203-1212, 1991.

Shimizu, Y., Van Seventer, G.A., Horgan, K.J., Shaw, S. Roles of adhesion molecules in T-cell recognition: Fundamental similarities between four integrins on resting human T cells (LFA-1, VLA-4, VLA-5, VLA-6) in expression, binding, and costimulation. *Immunol. Rev.* **114**:109-143, 1990.

Smith, W.B., Gamble, J.R., Clark-Lewis, I., Vadas, M.A. Interleukin-8 induces neutrophil transendothelial migration. *Immunology* , **72**:65-72, 1991.

Smith, C.W., Kishimoto, T.K., Abbass, O., Hughes, B., Rothlein, R., McIntire, L.V., Butcher, E.C., and Anderson, D.C. Chemotactic factors regulate lectin adhesion molecule 1 (LECAM-1)-dependent neutrophil adhesion to cytokine-stimulated endothelial cells *in vitro*. *J. Clin. Invest.* **87**:609-618, 1991.

Spertini, O., Kansas, G.S., Munro, J.M., Griffin, J.D., Tedder, T.F. Regulation of leukocyte migration by activation of the leukocyte adhesion molecule-1 (LAM-1) selectin. *Nature* **349**:691-694, 1991.

Springer T.A. Adhesion receptors of the immune system. *Nature* **346**:425-433, 1990.

Vedder, N.B., and Harlan, J.M. Increased surface expression of CD11/b/CD18 (Mac-1) is not required for stimulated neutrophil adherence to cultured endothelium. *J. Clin. Invest.* **81**:676-682, 1988.

von Andrian, U.H., Chambers, J.D., McEvoy, L., Bargatze, R.F., Arfors, K.-E., and Butcher, E.C. A two step model of leukocyte-endothelial cell interaction in inflammation: Distinct roles for LECAM-1 and the leukocyte ß2 integrins *in vivo*. *PNAS* **88**:7538-7542, 1991.

Walz, G., Aruffo, A., Kolanus, W., Bevilacqua, M., and Seed, B. Recognition by ELAM-1 of the sialyl Lewisx determinant of myeloid and tumor cells. Science **250**: 1132-1135, 1990.

Zimmerman, G.A., McIntyre, T.M., Mehra, M., and Prescott, S.M. Endothelial cell-associated platelet-activating factor: a novel mechanism of signaling intercellular adhesion. *J. Cell. Biol.* **110**:529-540, 1990.

CONTRIBUTORS

M. AMIN ARNAOUT--Department of Medicine, Massachusetts General Hospital, Harvard Medical School, Charlestown, Massachusetts 02114

EUGENE C. BUTCHER--Department of Pathology, Stanford University School of Medicine, Stanford, California 94305-5324

KENNETH DORSHKIND--Division of Biomedical Sciences, University of California, Riverside, California 92521-0121

BRIAN D. EVAVOLD--Department of Pathology, Washington University School of Medicine, St. Louis, Missouri 63110

SUDHIR GUPTA--Department of Medicine, University of California, Irvine, California 92717

MARTIN E. HEMLER--Divison of Tumor Virology, Dana-Farber Cancer Institute, Boston, Massachusetts 02115

MARILYN KEHRY--Department of Molecular Biology, Boehringer Ingelheim Pharmaceuticals, Ridgefield, Connecticut 06877

RICHARD D. KLAUSNER--Cell Biology and Metabolism Branch, NICHD, National Institutes of Health, Bethesda, Maryland 20892

ADA M. KRUISBEEK--Division of Immunology, Netherlands Cancer Institute, Amsterdam, The Netherlands

JEFFREY A. LEDBETTER--Bristol-Myers Squibb Pharmaceutical Research Institute, Seattle, Washington 98121

TAK W. MAK--Departments of Medical Biophysics and Immunology, The Ontario Cancer Institute, Toronto, Canada M4X 1K9

FRITZ MELCHERS--Basel Institute for Immunology, Basel, Switzerland

RANDOLPH J. NOELLE--Department of Microbiology, Dartmouth Medical College, Hanover, New Hampshire 03756

LINDA S. PARK--Department of Biochemistry, Immunex Research and Development Corporation, Seattle, Washington 98101

JANE R. PARNES--Department of Medicine, Stanford University School of Medicine, Stanford, California 94305

DAVID H. RAULET--Department of Molecular and Cell Biology, University of California, Berkeley, California 94720

LAWRENCE E. SAMELSON--Cell Biology and Metabolism Branch, NICHD, National Institutes of Health, Bethesda, Maryland 20892

STEPHEN SHAW--Experimental Immunology Branch, National Cancer Institute, National Institutes of Health, Bethesda, Maryland 20892

ETHAN M. SHEVACH--Laboratory of Immunology, NIAID, National Institutes of Health, Bethesda, Maryland 20892

IAN S. TROWBRIDGE--Department of Cancer Biology, The Salk Institute, La Jolla, California 92037

NICOLAI S.C. van OERS--Department of Microbiology, University of British Columbia, Vancouver, British Columbia, Canada V6T 1Z3

THOMAS A. WALDMANN--Metabolism Branch, National Cancer Institute, National Institutes of Health, Bethesda, Maryland 20892

THOMAS J. WALDSCHMIDT--Department of Pathology, University of Iowa, Iowa City, Iowa 52242

INDEX

CD28 receptor (continued)
  PLC γ1 and, 23-27
  tolerance induction and, 103
CD29 receptor, 182, 183
CD32 receptor, 151, 152-153, 155
CD44 receptor, 158, 184, 186
CD54 receptor, 172, 173, 174
CD62 receptor, 182
CD45RO cells, 159
Cell adhesion, 163-168, 171, 175
  CD31 cells in, 157-160
  leukocyte-endothelial, 181-190
Chemotherapy, 57
Chloroquine, 39
Chronic myelogenous leukemia, 59
c-kit proto-oncogene, 112, 121
Clonal deletions, 101, 102, 103, 105
Clonal inactivation, 101
Coated pit internalization, 175
Collagen, 76, 167, 168
Colony-stimulating factor (CSF), 120
Concanavalin A, 69
Crosslinking, 142-143, 153, 154, 159
  of CD28 receptor, 23-27
CSF, see Colony-stimulating factor
CTL, see Cytotoxic T cells
Cyclic adenosine monophosphate
            (cAMP), 135
Cycloheximide, 40, 41
Cyclosporin A, 141
  Pgp and, 39-46
Cyclosporine, 23
Cysteine, 4, 5, 33, 171-172
Cystic fibrosis gene, 45
Cytochrome C, 104
Cytokines, see also specific types
  B cell production and, 119-121
  CD28 receptor and, 23
  leukocyte-endothelial cell
            adhesion and, 189
  TCR γδ and, 49-50, 51, 53, 54
Cytoplasmic tails
  of CD18, 173-175
  of ζ dimers, 2
  of VLA-4, 163, 165-168
Cytoskeletal proteins, 176
Cytotoxic T cells, 2, 30, 46, 73, 75, 79
  tolerance induction and, 103-105

Diabetes, 76
ζ Dimers, 1-3, 10, 12, 13, 21
DNA, 75, 152

DNA synthesis, 141, 142, 143-145, 147
Drosophilia, 29, 33

ELAM, 182-183
Encephalomyelitis, 76
Endothelium, adhesion to, 157, 159,
            160, 175
  leukocyte, 181-190
Eosinophils, 163
Erythropoietin, 58
E-selectin, 182-183, 189
Ezrin, 11, 33

Factor X, 172
Fc receptors, 3, 11
Fcε RII receptor, see CD23 receptor
Fcγ RII receptor, see CD32 receptor
Fibrinogen, 50, 172
Fibroblasts, 11
Fibronectin, 50, 158, 163, 164, 167-
            168
FMLP, 173
FN-40, 164, 166

G-CSF, see Granulocyte colony-
            stimulating factor
Genistein, 135
Glue molecules, 158
Glutamic acid, 18, 19, 20, 164
Glutamine synthetase, 30
Glycine, 18
Glycophosphatidyl inositol (GPI), 3, 13
GM-CSF, see Granulocyte-macrophage
            colony-stimulating factor
GMP140, 182
GPIIb-IIIa, 189
GPI, see Glycophosphatidyl inositol
G proteins, 173
Graft-versus-host disease, 63
Granulocyte colony-stimulating factor
            (G-CSF), 58
Granulocyte-macrophage colony-
            stimulating factor (GM-
            CSF), 51, 58, 112
Gross murine leukemia virus, 103, 105
Growth factor receptors, 10, 11, 175
Growth hormone, 58

Hb(64-76) determinant, 17-21
Heat shock proteins, 49, 54
Heat stable antigen (HSA), 127, 128,
            129

ε Heavy chains, 150, 154
γ Heavy chains, 150
μ Heavy chains, 113-115, 119
HEBFpp receptor, 184
Helper T cells, 73, 75
    B cells and, 131-137, 139-147
    CD45 glycoprotein and, 30
    CD28 receptor and, 26
    Tac and, 59
Hematopoietic cells, 12, 68, 111, 119,
        120
Hemoglobin, 17
Herbimycin A, 26
Histiocytic lymphoma, 59
HIV-1, see Human immunodeficiency
        virus 1
Hodgkin's disease, 59
HSA, see Heat stable antigen
HTLV-1, see Human T lymphocyte
        virus-1
Human immunodeficiency virus-1 (HIV-
        1), 39
Human T lymphocyte virus-1 (HTLV-
        1), 59-61
HuMIP-1ß, 186
Humoral immunity, 131
Hybrid CD8/CD4 transgene, 79-85
Hydroxyurea, 39-40

ICAM, see Intercellular adhesion
        molecule
iC3b, 172, 173, 174
IFN, see Interferon
Ig, see Immunoglobulin
IL, see Interleukin
Immunoglobulin (Ig), 112, 113, 114,
        115, 119, 120, 121, see
        also specific types
    CD7E-, 133, 134
    CD23 receptor and, 153
    helper cells and, 132, 133, 134,
        144
Immunoglobulin A (IgA), 144
Immunoglobulin D (IgD), 128, 129,
        135, 139, 153, see also
        Anti-immunoglobulin D
        monoclonal antibodies
Immunoglobulin E (IgE), 11, 149-155,
        see also CD23 receptor
    helper cells and, 134, 144, 145-
        146, 147
Immunoglobulin G (IgG), 61

Immunoglobulin G (IgG) (continued)
    CD23 receptor and, 149, 150, 151,
        153
    helper cells and, 133, 137, 144,
        145-146
Immunoglobulin H (IgH), 115
Immunoglobulin M (IgM), 113, 114,
        126, 127, 128, 129, 139
    helper cells and, 135, 137, 141,
        142, 144, 145-146
Inositol phosphates, 23
Insulin receptor, 175
Integrins, 158, 159, see also ß Integrins
    TCR γδ and, 49-55
ß1 Integrins, 185
ß2 Integrins
    leukocyte-endothelial cell adhesion
        and, 182, 183, 184-185
    regulation of, 171-176
Intercellular adhesion molecule-1
        (ICAM-1), 59, 131, 132,
        141, 158, 159, 160, 185
Intercellular adhesion molecule-2
        (ICAM-2), 158, 160, 172
Intercellular adhesion molecule-3
        (ICAM-3), 158, 160
Interferon-γ (IFN-γ), 141, 144
Interleukin-1 (IL-1), 20, 144, 186
Interleukin-2 (IL-2), 141, 144, see also
        Interleukin-2 receptor
    CD28 receptor and, 23, 25, 26
    TCR γδ and, 49, 51, 52, 53
Interleukin-3 (IL-3), 51, 58, 141
Interleukin-4 (IL-4), 58, 121, 132, 134,
        136-137, 141, 145, 147
    CD23 receptor and, 153
    Hb(64-76) and, 19-21
    TCR γδ and, 50, 51
Interleukin-5 (IL-5), 132, 137, 141, 145
Interleukin-6 (IL-6), 141, 144
Interleukin-7 (IL-7), 111, 112-114, 115,
        120-121, 125-129
    in vivo administration of, 125-126
Interleukin-B (IL-B), 184
Interleukin-2 receptor (Tac), 57-63
    in autoimmune disorders, 59
    in cancer, 59
    structure and function of, 58-59
    TCR and, 2-3
Ionomycin, 23, 26, 32
Isoleucine, 18

Monoclonal antibodies (continued)
    anti-CD28, 23, 25-26
    anti-CD31, 159
    anti-CD40, 133
    anti-CD45, 31
    anti-IgE, 150-151
    anti-IgM, 149
    anti-ß2 integrin, 185
    anti-phosphotyrosine, 10, 30, 94
    anti-Tac, 57-58, 60-61, 62-63
    anti-TCR, 50
    anti-TCR αß, 91
    B3B4, 154
    IgD, see Anti-immunoglobulin D
        monoclonal antibodies
    MR1, 134
    MRK-16, 39, 40, 43, 45
    NKI-L16, 173
    Pgp and, 39, 40, 43, 45
Monocytes, 59, 159, 163, 185
Mouse mammary tumor virus (MMTV),
        102, 103, 105
MRK-16 monoclonal antibodies, 39, 40,
        43, 45
MRI monoclonal antibodies, 134
mRNA, 141
    CD45, 32, 33
    c-kit, 121
    cytokine, 23
    IL2, 23, 25, 26
    mdr 1, 45, 46
Multidrug resistant (mdr 1) gene, 39-46
Murine lymphoma cell lines, 30

Natural killer (NK) cells
    ß2 integrins and, 171
    MHC class I and, 67, 69
Negative selection, 89, 90, 101, 105
Neomycin resistance gene, 76
Neonatal tolerance induction, 101-102,
        105
Neutrophilia, 172
Neutrophils
    ß2 integrins and, 172, 173, 175
    leukocyte-endothelial cell
        adhesion and, 182-185
NK cells, see Natural killer cells
NKI-L16, 174
NKI-L16 monoclonal antibodies, 173

Orthovanadate, 10, 11
Ovalbumin, 102

PADGEM, 182
Peptides, 140
    tolerance induction and, 101-
        102, 103-106
p59+*fyn*+ protein tyrosine kinase
        (PTK), 27
    T cell development and, 89-97
p60+*fyn*+ protein tyrosine kinase
        (PTK), 12
P-glycoprotein (Pgp), 39-46
Pgp, see P-glycoprotein
PHA, see Phytohemagglutinin
Phenylalanine, 18-19
Phenylarsine oxide, 10, 11
Phorbol-12-myristate-13-acetate (PMA)
    CD28 receptor and, 23, 25, 26
    Pgp and, 40-41, 42
    TCR γδ and, 52, 53, 54
Phospholipase C, 90, see also
        Phospolipase C γ1
Phospholipase C γ1 (PLC γ1), 1, 10
    CD28 receptor and, 23-27
Phycoerythrin, 80
Phytohemagglutinin (PHA)
    CD28 receptor and, 24, 25, 26
    Pgp and, 39-41, 42, 44, 45, 46
PKA, see Protein kinase A
PKC, see Protein kinase C
Plasmodium falciparum infections, 39
PLC, see Phospholipase C
p56+*lck*+ protein tyrosine kinase
        (PTK), 12, 27, 30
    T cell development and, 89-97
PMA, see Phorbol-12-myristate-13-
        acetate
PM+ACT+, 132, 133, 135, 136, 136-
        137
PM+REST+, 137
Polyclonal antibodies, 91, 94, 134
Positive selection, 90, 101
p55 peptides, 58
p75 peptides, 58
p81 proteins, 10
P100 proteins, 10
Prolactin, 58
Protein kinase A (PKA), 173
Protein kinase C (PKC), 135, 173
    Pgp and, 39-46
Protein tyrosine kinase (PTK), 9-14, 29,
        see also specific types
    characterization of substrates in,
        10-11

The manufacturer's authorised representative in the EU is Springer
Nature Customer Service Centre GmbH, Europaplatz 3, 69115 Heidelberg,
Germany. If you have any concerns regarding our products, please
contact ProductSafety@springernature.com

Printed and bound by CPI Group (UK) Ltd, Croydon, CR0 4YY
23/04/2026
02095624-0011